Water Policy Processes in India

The privatization of water is a keenly contested issue in an economically liberalizing India. Since the 1990s, large social groups across India's diverse and disparate peoples have been re-negotiating their cultural relationships with each other as to whether they support or oppose pro-privatization water policy reforms. These claims and counter claims are seen as an impending war over water resources, one that includes many different players, located across a wide variety of sites with many different agendas, whose actions and interactions shape policy production in India.

This book is one of the first to assess the dynamics of urban water policy processes in India. Using the case study of Delhi's water situation, this book analyzes emergent dynamics of policy process in India in general and, more specifically, in the post-economic reform era. Taking as its starting point a critique of linear versions of policy making, the author explains both how and why particular types of knowledge, practices and values get established in policy as well as the complex interplay of knowledge, power and agency in water policy processes.

Water Policy Processes in India covers a critical gap in the literature by analyzing how governments in practice make policies that greatly affect the welfare of their people; analyzing the process through which policies are developed and implemented; investigating the aims and motives behind policies; and identifying the potential areas of intervention in order to improve the policy process in both its development and implementation stages.

Vandana Asthana teaches international politics in the Department of Government at Eastern Washington University, USA. She is a founder member of the IC Centre for Governance, New Delhi, and the founder secretary and member of the advisory panel of Eco-Friends, an NGO working on water issues in India. She has served as a consultant and completed a project for the Government of India on water security and is a board member of the South Asian Studies Association.

Routledge Contemporary South Asia Series

1 **Pakistan**
Social and Cultural
Transformations in a Muslim
Nation
Mohammad A. Qadeer

2 **Labor, Democratization and
Development in India and
Pakistan**
Christopher Candland

3 **China-India Relations**
Contemporary dynamics
Amardeep Athwal

4 **Madrasas in South Asia**
Teaching terror?
Jamal Malik

5 **Labor, Globalization and the
State**
Workers, women and migrants
confront neoliberalism
*Edited by Debdas Banerjee and
Michael Goldfield*

6 **Indian Literature and Popular
Cinema**
Recasting classics
Edited by Heidi R.M. Pauwels

7 **Islamist Militancy in Bangladesh**
A complex web
Ali Riaz

8 **Regionalism in South Asia**
Negotiating cooperation, institu-
tional structures
Kishore C. Dash

9 **Federalism, Nationalism and
Development**
India and the Punjab Economy
Pritam Singh

10 **Human Development and Social
Power**
Perspectives from South Asia
Ananya Mukherjee Reed

11 **The South Asian Diaspora**
Transnational networks and
changing identities
*Edited by Rajesh Rai and Peter
Reeves*

12 **Pakistan-Japan Relations**
Continuity and change in economic relations and security
interests
Ahmad Rashid Malik

13 **Himalayan Frontiers of India**
Historical, Geo-Political and
Strategic Perspectives
K. Warikoo

14 **India's Open-Economy Policy**
Globalism, Rivalry, Continuity
Jalal Alamgir

15 **The Separatist Conflict in Sri
Lanka**
Terrorism, Ethnicity, Political
Economy
Asoka Bandarage

16 **India's Energy Security**
*Edited by Ligia Noronha and
Anant Sudarshan*

17 **Globalization and the Middle
Classes in India**
The Social and Cultural Impact of
Neoliberal Reforms
*Ruchira Ganguly-Scrase and
Timothy J. Scrase*

18 **Water Policy Processes in India**
Discourses of Power and
Resistance
Vandana Asthana

Water Policy Processes in India

Discourses of power and resistance

Vandana Asthana

LONDON AND NEW YORK

First published 2009
by Routledge
2 Park Square, Milton Park, Abingdon, Oxon OX14 4RN

Simultaneously published in the USA and Canada
by Routledge
711 Third Ave, New York, NY 10017

Routledge is an imprint of the Taylor & Francis Group, an Informa business

© 2009 Vandana Asthana

First issued in paperback 2012

Typeset in Times New Roman by Pindar NZ, Auckland, New Zealand

British Library Cataloguing in Publication Data
A catalogue record for this book is available from the British Library

Library of Congress Cataloging-in-Publication Data
Water policy processes in India: discourses of power and resistance/
Vandana Asthana.
       p. cm.—(Routledge contemporary South Asia series; 18)
    Includes bibliographical references and index.
    1. Water supply—Government policy—India. 2. Water resources
development—Inida. I. Title.
    HD1698.I4A88 2009
    33.9100954—dc22   2009002243

ISBN 13: 978-0-415-77831-2 (hbk)
ISBN 13: 978-0-203-87421-9 (ebk)
ISBN 13: 978-0-415-32740-5 (pbk)

Contents

List of illustrations   viii
Preface   x
List of abbreviations   xiii

1   Introduction   1

2   Changing the frame: repositioning policy   9

3   The process of economic liberalization and private sector
    participation   29

4   Water in the liberalization process   45

5   Situating Delhi in the water reform project   71

6   Mainstreaming policy: discourses of power   99

7   Creating spaces for change: collective action on the water
    reform project   117

8   Understanding the water policy process   139

    References   154
    Index   168

Illustrations

Figures

2.1	Conceptual framework	26
3.1	Change team: ideas and power model (1985–2004)	34
5.1	Location of Delhi in India	73
5.2	Existing and proposed water treatment plants	76
5.3	Demand, supply and delivery projections in millions of liters per day (MLD)	78
5.4	Inequality in Delhi water supply	80
5.5	Map of water augmentation route from the Tehri Dam to the Sonia Vihar Plant, Delhi	83
5.6	Augmentation of water from the Ganga River near Old Tehri Town	84
5.7	The Tehri Dam under construction	84
5.8	The Ganga Canal at Hardwar	85
5.9	NBCC pipes	86
5.10	Pipeline carrying water to Delhi	86
5.11	Location of the Sonia Vihar Plant in Delhi	87
8.1	Actors in the Delhi water reform project	139
8.2	Networks of power	144
8.3	Networks of resistance	145
8.4	Linear and horizontal models of policy making	149
8.5	Policy production in the post-reform era	151

Tables

4.1	World Bank reports on India's water sector: an overview	56

5.1   Delhi's water supply capacity in millions of gallons per day
      (MGD)   78
5.2   Percentage of population with inadequate water supply in Delhi   80
5.3   Nonrevenue water in Delhi   81
5.4   Proposals for augmentation of water supply   83
5.5   Chronology of major events in the Delhi Water Supply and
      Sewerage Sector Reform Project (DWSSSRP)   94

Preface

Knowledge of public policy is socially relevant. It forms an important agenda for the democratic functioning of a nation and plays a decisive role in the destiny of any country. With over a billion people, India is the largest functioning democracy in the world, with a vibrant civil society, a free press and the freedom to debate the rationale of government policies that impinge upon the daily lives of citizens in countless ways. In order to understand how governments make policies that so greatly affect the welfare of their people, it is necessary to examine how the policy making process works in practice, rather than simply in theory. *Water Policy Processes in India: Discourses of power and resistance* reveals the making of policy through the lens of water in the post-economic reform era, a process that includes many different players with many different agendas located across a wide variety of sites. This book is an effort to systematically analyze the practicality of policy processes in India through a detailed focus on water issues in the face of rhetoric, polemics and activism.

The privatization of water is a keenly contested issue in an economically liberalizing India. Since the 1990s, large social groups across India's diverse and disparate peoples are re-negotiating their cultural relationships with each other according to whether they support or oppose pro-privatization water policy reforms. Through an ethnographic study of Delhi's water situation, and in contrast to policy impact studies, this book looks at policy production in a neoliberal era. It examines how water policy is produced in relation to local, subnational, national, and global pressures, radiating from "above" and "below" the state of Delhi. This volume is not only important for India and the region, where processes of globalization and neoclassical economics are dominating government planning and development issues – it is also important for understanding how processes of globalization function in the developing world.

The strength of the book lies, firstly, in the specific area of policy process in

India, hitherto understudied. Since policy processes include not only macro policy development but also its implementation and interpretation by different agencies, there is a need to relate policy to the institutional and social processes through which it is formed, translated, and implemented, and to identify where critical gaps exist in current texts on policy analysis. The book aims to fill this gap in the literature by analyzing the process through which policies are developed and implemented; investigating the aims and motives behind policies; and identifying the potential areas of intervention in order to improve the policy process in both its development and implementation stages. Secondly, while much has been written about policy making in India, relatively little attention has been focused on the emergent dynamics of policy process in India in general and more specifically in water – a body of literature this volume significantly enhances.

Whereas in previous eras state support and plan budgets were allocated by the central Government of India, and urban water provisions, management and delivery were the responsibility of individual states, in today's world water has become part of the infrastructure sector and its reform is considered essential for growth and development. Water thus provides an entry point from which to explore complex and dynamic policy processes with the competing narratives and diverse actors of the post-reform era. This book provides an introduction to the policy process literature in general and in India in particular; the dynamics of the process of economic liberalization and urban water policy in the liberalization process; the water policy reform process in Delhi and the discourses of power and resistance in relation to policy making and understanding policy processes. This sets the book apart and places it in its own scholarly niche, providing a new dimension to public policy process and water in an age of neoliberalism and globalization.

Water Policy Processes in India is intended for scholars in the interdisciplinary subjects of public policy and environment, as well as readers with an interest in water policy in India and the region. It will be of interest to bureaucrats and policy planners, scholars and nongovernmental organizations (NGOs) working on water issues, and should find a place in their personal and institutional libraries.

A word of thanks is due to all those people and institutions whose support, knowledge and input form an essential part of the story on water policy processes. The author is grateful to the authors of books and research articles on public policy, scholars, NGOs, civil society members, religious and political leaders, administrators, bureaucrats and water stakeholders whose views, expressed in personal interviews, are incorporated in the themes of this book. A special thanks goes to Dr Vandana Shiva and the members of the Research Foundation for Science, Technology and Ecology who were of immense assistance with my research and facilitated my access to networks of the Water Liberation Campaign and the Citizens Front for Water Democracy. The author also acknowledges the Nehru Memorial library, New Delhi, the Centre for Science and Environment Library, New Delhi, and the libraries of the University of Illinois at Urbana Champaign and the Eastern Washington University at Cheney for providing access to a wide body of literature. Finally, thanks are due to authors of the works cited and referenced herein and of all the other sources acknowledged in the text.

Abbreviations

ADB	Asian Development Bank
BJP	Bhartiya Janata Party
BMZ	Federal Ministry for Economic Cooperation and Development (Germany)
BOO	Build-Own-Operate
BOT	Build-Operate-Transfer
CII	Confederation of Indian Industry
CPI (M)	Communist Party of India (Marxist)
CFWD	Citizens Front for Water Democracy
DCB	Delhi Cantonment Board
DFID	Department for International Development
DJB	Delhi Jal Board
DMAs	District-metering areas
DUEIIP	Delhi Urban Environment and Infrastructure Improvement Project
DWSSSRP	Delhi Water Supply and Sewerage Sector Reform Project
GATS	General Agreement on Trade in Services
GATT	General Agreement on Trade and Tariffs
IBAW	Indian Business Alliance on Water
IDS	Institute of Development Studies
IFI	International Financial Institutions
IGOs	Intergovernmental organizations
IIM	Indian Institute of Management
IIT	Indian Institute of Technology
ILFS	Infrastructure Leasing and Financial Services
IMF	International Monetary Fund
MGD	Million gallons per day

MNC	Multinational corporation
MOWR	Ministry of Water Resources
NAC	National Advisory Council
NCT	National Capital Territory
NDMC	New Delhi Municipal Council
NGO	Nongovernmental organization
NPM	New public management
NWP	National Water Policy
O&M	Operation and maintenance
ODI	Overseas Development Institute
OECF	Overseas Economic Cooperation Fund
OZ	Operational Zones
PPIAF	Public–Private Infrastructure Advisory Facility
PPP	Public–private partnership
PSE	Public sector undertakings
PwC	PricewaterhouseCoopers
RFSTE	Research Foundation for Science Technology and Ecology
RWA	Resident Water Association
SAIN	Sain International
SEB	State Electricity Board
ULB	Urban local bodies
UNDP	United Nation Development Program
UNICEF	United Nations International Children Emergency Fund
USAID	United States Agency for International Development
WB	World Bank
WEF	World Economic Forum
WHO	World Health Organization
WLC	Water Liberation Campaign
WRSS	World Bank Water Resources Sector Strategy
WTO	World Trade Organization
WWA	Water Workers Alliance

1 Introduction

This book is a story of the making of a policy, one that could not be more vital to the people whose lives policy making touches, and one that includes many different players with many different agendas located across a wide variety of sites. This is a book about policy making processes, an understanding of which is fundamental to a full appreciation of the impact that policies have on everyone's daily lives. *Water Policy Processes in India* analyzes the policy making process in the post economic reform era of India beginning 1991. The argument made here is that since the 1990s globalization and the rescaling of the state have brought about a contextual "messiness" in the debate about policy production. Viewed in relation to global pressures, as well as local, subnational, and national imperatives, this volume addresses the contemporary shifts in policy production in post-reform India that can be seen as a microcosm of the policy making process in all its complexity across the developing world. By (1) analyzing the process through which policies are developed and implemented, (2) investigating the aims and motives behind policies, and (3) identifying the potential areas of intervention in order to improve the policy process in both its development and implementation stages, this book attempts to understand what policy making means in practice, and in all its complexities brought about by the processes of globalization.

The subject of policy making processes is relatively understudied in most developing societies, including India. The areas that need attention and studies in the Indian policy process include (1) resource disparities; (2) uneven representation in governance; (3) the role of knowledge power networks in reproducing inequality in the policy process by delegitimizing local knowledge; (4) the context of policy discourse – the nationalist central planning framework of the post-independence period and the influence of global discourses more recently; and (5) the institutional relations – the dominance of state institutions in the early post-independence period and the swing toward market mechanisms in the 1990s (Arora 2002). This volume dwells within paradigms of most of these concerns and, through the lens of water policy processes, seeks to determine whether the new economic and political dispensation in India means a new politics of policy making.

The classical model of policy making views it as a linear process: it assumes that policies are the direct result of a rational process of problem identification by a benevolent agency (usually the state) which then "prescribes" the appropriate solution. *Water Policy Processes in India* questions some of these assumptions. In place of the purely linear model, it offers a multilayered, multivalent scheme in which power and influence move in many different directions and flow from many different sources.

Unlike the "policy as prescription" approach, which characterizes much of the public administration literature and has remained popular with policy makers, the approach of this book reflects a fast expanding body of literature that analyzes, explains or conceptualizes the process dimensions of policy. It does not assume that policies are "natural phenomena" or "automatic solutions" resulting from particular social problems and it does not privilege the state as an actor fundamentally different from other social actors. Questions of causation, method, and agency are treated empirically, and answers are derived from concrete empirical research designed to encompass the complexities of the policy making process in an age of globalization. But first we must turn to understand the processes of globalization and the effect they have on state and policy making.

Globalization, the state and policy production

The term *globalization* was first used in 1985 by Theodore Levitt (1985) to characterize vast changes in the international economy – changes related to the rapid, pervasive and global diffusion of the production, consumption, and investment of goods, services, capital, and technology. Initially used by economic historians, this term is now increasingly used in social science literature to describe a variety of economic, political, social, and cultural changes (Finger and Allouche 2002: 2). Giddens, for example, reiterated its importance in social sciences and defined globalization as a process leading to the "intensification" of worldwide social relations which link distant localities in such a way that local happenings are shaped by events occurring many miles away and vice versa (Giddens 1990: 64).

The interconnectedness of various social phenomena around the globe is also stressed by McGrew and Lewis (1992: 22) who define globalization as "the process by which events, decisions, and activities in one part of the world come to have significant consequences for individuals and communities in quite distant parts of the globe."

In the literature of political science and international relations, globalization usually points to the post-Westphalian era (e.g. Held 1995) or as Rosenau (1997) sees it, to a "post-international" system. Thus, for political scientists and international relations specialists, globalization defines both a process and a situation in which political relationships are less territorially based, and nation-states become less important. As a part of this process, decision making power is gradually removed from the nation-states and shifted to other actors, which can be located "above," "below," and "beside" the nation-states. This repositioning has a dual effect. While technological change and economic integration have pushed the state in the direction of greater conformity and adoption of global standards and behavior, social and political forces have pulled in the direction of asserting the state's role and power in protecting the interests and the livelihoods of its citizens.

Literature on globalization reveals that the role of the state has changed in policy making processes. The state is no longer the sole authority for protecting the interests of the poor and disadvantaged; other social, economic and political institutions are emerging as players in this process of governance. Commonly referred to as a "multiplication of all kinds of governance," this development brings into play a constellation of actors including various institutions of the state apparatus, intergovernmental organizations (IGOs), social movements, and local actors (Rosenau 2002: 230). Many global process analysts hold that the state is becoming increasingly insignificant in policy making. The retreat and erosion (Strange 1996), "the hollowing out" (Jessop 1999), and the changing architecture (Cerny 1990) of the state in response to global pressures are typical concerns. Many believe that it is the end of the nation state and a forced retreat of the welfare state (Ohmae 1995). While "weak state" theories in globalization literature point to the *erosion* of the state, other accounts point to the *transformation* of the state, arguing that it still maintains a highly significant and strategic role (Sassen 1996). States may cede authority over some aspects of economic and financial governance but there are areas where states retain a substantial degree of control.

In her book *The Myth of a Powerless State*, Linda Weiss (1998) offers a contemporary critique of the popular feeling among analysts that globalization erodes the power of the nation-state. She contends that the "end-of-the-state literature" exaggerates the powers the state had in the past, overlooks the diversity of the capacity of the state, and politically constructs the notion of helplessness. It is the transformative capacity of the state, Weiss believes, that can give it a competitive edge in the global economy.

Whatever be the positions of various analysts, all the extensive body of scholarly work on globalization indicates that the state is under pressure. Financial and economic globalization, along with deregulation and the opening up of government to market forces, have weakened the ability of the state to conduct macroeconomic policy. With competition emerging in many sectors, including that of infrastructure, states are compelled to loosen control over many of their public sector operations. Another factor that has put pressure on the state has been the need to compete internationally, leading to high public debt that states need to address in order to adjust to the new global economy. According to Finger and Allouche (2002), this is one of the main reasons why states have undergone substantial transformation since the 1980s in both the North and the South. Another reason cited for the recent transformation of the state has been the "legitimation crisis" (Habermas 1980; Offe 1984). In an increasingly global economy and culture, the state is more and more challenged to legitimize itself in the light of pressures from "above" and "below."

It is therefore not surprising that states have started to adjust to this shift. While in the North this change takes the form of new public management (NPM) efforts, in the South it takes the form of structural adjustment programs mandated by the International Monetary Fund (IMF) and the World Bank. These adjustments have basically been in accordance with the neoclassical or neoliberal ideologies of the increased role of the market and the reduced importance of the state in almost all sectors, but especially so in service delivery (Finger and Allouche 2002). Such ideologically motivated reforms tend to treat the state as a factor upsetting the ideal optimum equilibrium of the market's "invisible hand" in the developmental process. Under this approach the role of the corporate sector in growth and development is to provide public good to the society. Issues involving land, water, and forests are seen as management problems which can be more efficiently handled by corporations than by the state; hence, the market in the neoliberal world will supposedly benefit society. Although this ideology has not been without its critics, it remains the dominant paradigm in governance and policy making in the state. The state in the age of globalization thus finds its policy making and governance functions involved in a constant interaction between external pressures from "above" and internal pressures from "below" and the Indian state has been no exception to this shift. The following section provides background to the structure that the Indian state has adopted in the neoliberal era.

Policy production and the Indian state

After independence India decided to follow the model provided by the British colonial administrative system. It was a decision that resulted in the massive expansion of a largely unreconstructed colonial bureaucracy. The bureaucracy of the newly independent Indian state remained highly centralized, largely working on a 'command and control' administrative style in the areas of both policy and planning (Kaviraj 1997: 233–4).

In the politically unstable situation that existed during the independence era after the partition of the country, the new leadership felt that a strong state needed to be established in order to maintain a sense of national unity. This resolve was clearly recognized between 1947 and 1950, when the Indian constitution was drafted. Consequently, within the framework of a federal system, the constitution gave strong political and economic powers to the center. Of particular importance was the power to allocate financial resources between the center and individual states and the center's support for state projects within the social sector in areas such as irrigation, water supplies, etc.

However, in response to the new exigencies of economic liberalization which began in 1991, the Indian state has undergone major deviations from the way it had traditionally conducted its centralized policy planning since the post-independence era. The central Government delegated more power and more responsibility to the individual states yet reduced the amount of money available to carry out necessary programs. This decrease in state support and centrally directed state planning has reduced the fiscal independence of the state Governments and increased the importance of attracting foreign direct investment. With funds curtailed by central Government policies, many states started to seek alternative sources in the form of loans and grants from international agencies. The decrease in state support and funding cuts has meant that other agencies like the private sector and donors are acquiring a new significance. These factors have led to a new kind of policy making and raises important questions of public policy in the liberalization era.

The choice of water

Since the focus of the book is policy production in the post-economic reform era, water provided the conceptual lens through which to understand these processes because over the past two decades a remarkable transformation has taken place in Indian attitudes towards water. After a period of state dominance for much of the twentieth century, water management is undergoing a dramatic transformation through the process of privatization, liberalization and deregulation. This new policy framework is built upon a set of principles essentially imported from Washington – the principles of liberalization, privatization, the free market, and structural adjustment. Through an understanding with the State Department of the USA, the IMF and the World Bank, these principles have come to dominate the political and bureaucratic corridors of India. India began to seek its own space in the global paradigm shift by liberalizing and privatizing its public and social sector to fit into the existing system of international trade, technology and the capital revolve cycle. Although this shift to market orientation in the area of water policy retains unquestioned popularity among the states, it has been hotly debated and contested by activists and nongovernmental organizations (NGOs). This book contributes to the literature on the post-reform era and demonstrates how economic reforms and their political and economic context have led to accommodations at the state level bringing a new politics of policy making to India. The number of actors has increased in the post-reform era, and networks of actors exist both inside and outside the state. The form and location of expertise and sites of policy making are multiscalar and debate and framings overlap at all levels.

Another reason why water was an appropriate choice can be explained by the literature available on water. Mollinga (2005) places the literature on water into three categories: (1) literature on desirable water policy, (2) empirical research on water resource management, and (3) critical, oppositional discourse. All of these categories deal with substantial scholarly research in policy and research for policy, but, as Mollinga points out, there has been very limited research on policy processes. His analysis shows that there has been very limited research on policy production analysis from the standpoint of water and that there is a lack of literature on water resource management policy processes in India (Mollinga (2005: 7–14). Ramaswamy Iyer's book *Water Perspectives* (2003) describes, among other issues, the history of certain administrative and policy processes in the water sector which can be used for a background context for further inquiry into the dynamics of water policy processes. However, Mollinga (2005:12) argues that:

> We don't know very much about what happens in the water policy domain: how new policies are articulated, how the hydrocracies precisely work, how policy elites and their networks operate, how lobby and agitation is dealt with by the government administration, how the negotiations between government and international donors takes place, and so forth. One could ask – where are the political scientists in the water sector; who is looking in detail, as a researcher, at the internal dynamics of India's domestic hydropolitics?

Water policy production in the state of Delhi – the subject of this book – therefore provides an excellent example of the policy pressures and conflicting aims associated with the sorts of complex, multiscalar policy production processes that have resulted from globalization. While the predominant area of theorization in public policy and water governance has focused on concepts such as agenda setting of the state, there has been little theorization of the overlap of the local, subnational, national, and global scales, which may be expressed by a variety of voices, and the dynamics of inclusion and exclusion that surround the widely diverse actors involved in the processes of producing water policy. This book therefore frames shifts in policy production in a way that does not always revolve around the state but delves deeply into interactions between complex constellations of actors.

These state and non-state actors both use and are bound by various discourses of reform in urban water policy. Their competing interpretations are at the root of key debates over Delhi's water, which has enormous implications for how water policy is constructed, as well as the process that policy undergoes in its negotiations and implementation. In trying to understand the policy process as it emerges out of the competing interactions of the various actors, a mixed methods approach has been adopted that combines ethnography (Emerson, *et al.* 1995; Marcus 1995), and interview (Babbie 1998), to construct an analytical map of the practices and discourses that combine to form the unique policy configurations of Delhi's urban landscape.

The controversy over a contract awarded to multinational corporation (MNC) by the Government of Delhi for a water treatment plant in Sonia Vihar provides the empirical groundwork for this mapping process. The water treatment plant was to be managed by the French MNC Ondeo Degremont, a subsidiary of Suez and one of the biggest companies in the world after Vivendi in the area of water supply, distillation, and purification. The water from the plant was to provide water for a 24/7 pilot project which aimed to provide universal 24/7 safe water and sewerage services in an equitable, efficient and sustainable manner through a customer oriented and accountable service provider to the people of Delhi. However, the Government's water project was contested on the grounds that turning over management to this MNC would ultimately lead to the privatization of water in Delhi, which in turn would result in higher water tariffs and differential access to water for the populace. The questions driving this book came from an interest in competing values attached to water at multiple levels of society and in the lived experiences of the contestations and collaborations in which policies are negotiated, implemented, and reformed. The aim was to map the dialogical tensions in the production of water policy to understand the complex processes through which actors with power differences are attempting to manipulate access to water.

By drawing attention to the multiple ways in which water is "valued" in urban water reform in Delhi and how the discourse of valuation is used by actors to position themselves and their interests in Delhi's water management policies, this book raises questions regarding the nature of policy making in Delhi's water resource development and management practices in a neoliberal era. Rather than focusing on one pervasive discourse and studying its effect on people and their environment, the focus is on drawing attention to the roles of multiple actors, their history, and contestations that help expose the complexity and dynamics of water policy production.

To understand therefore, how polices are being made in practice, the importance of the relationships embedded as they are in complex networks of divergent and overlapping interests that create a particular political economy of knowledge and power needs to be understood. By tracing the origin of the water policy in the state of India, examining its historical, economic and political context and the key debates in relation to Delhi's engagement with global pressures, as well as pressures emanating from local, subnational, and national imperatives, one can begin to understand what policy making means in practice. And in a larger sense, this book seeks answers to the following questions: What is policy? What are the processes – technical, political, intellectual, and social in policy making? What do these inputs suggest about the nature of policy making in general and in a sector as important as water in particular, a sector seen as the key to economic growth and prosperity in the arena of development and central to the new economy era?

2 Changing the frame

Repositioning policy

This study of the dynamics of making and shaping urban water reform policy uses Delhi as a case study and takes as its starting point a critique of linear versions of policy making, highlighting the complex interplay of power, knowledge, and agency in water policy processes. Policy, as argued in this chapter, is not shaped simply on the basis of agenda setting preceded by good research and information, nor does it emerge from bargaining amongst actors on clearly defined options and choices. Rather, it is a more complex process through which particular interpretations of water come to frame what counts as knowledge and whose voices count in the deliberations in particular political and institutional contexts. To conceptualize policy as a process means that we need to explicitly acknowledge the importance of the social and historical context in which policy is shaped and implemented (Mooij 2003). This means that policy processes are likely to be contextual and vary across countries, political systems, and policy areas.

Against this backdrop, the literature on policy process is divided into four sections in this chapter. Section one begins with a review of general policy process literature. Section two goes on to discuss the literature available on policy processes in India. Section three specifically reviews literature on water policy processes in India. After a review of the literature, the last section concludes with a conceptual framework that includes the role of power, knowledge, and agency in the framing of water policy.

Literature on policy process

The literature on the conceptualization of policy process is represented by three broad approaches: (1) the linear model, (2) the actor network model, and (3) the power, interests and knowledge model. This section begins with a review of the linear policy approach and its critique, followed by the actor network approach where actors bargain with each other to produce a policy outcome. Lastly, it discusses the discourse approach, and how this shapes and guides policy problems and courses of action. It also focuses on the role of power and knowledge in structuring policy arrangements under the category of policy making as discourse.

Linear view of policy making

According to the traditional view, policy constitutes the decisions taken by those with responsibility for a given policy area, and these decisions usually take the form of statements or formal positions on an issue, which are then executed by the bureaucracy (Keeley and Scoones 1999; Considine 1994; Howlett and Ramesh 1995). The main assumption behind this concept is that "policy is a purposive course of action" (Anderson, *et al.* 1984: 4) and a "projected program of goals, values and practices" (Lasswell and Kaplan 1970: 71) which is to be decided by an authority that is coherent, instrumental, and hierarchical (Colebatch 1998: 43). Conceived in this way, policy is a product of a linear process moving through stages of agenda setting, decision making and finally implementation. From this perspective, policy making consists of four stages:

- **Agenda setting.** This is the stage in which goals are determined by the decisions of the authorized leaders who are supposed to act rationally.
- **Decision making.** Having clarified their goals, the leaders then select a course of action that will help realize them.
- **Implementation.** Other subordinates then have to implement the chosen course of action, and the rest of the organizational process in the institution is described as the implementation of these choices.
- **Evaluation.** The outcome of implementation of the decision is evaluated, and if necessary, policy may then be amended in light of the evaluation (Colebatch 1998: 43).

Because this linear presentation of policy making is based on the assumption of an orderly progression from objective to outcome, policy making is often seen as a series of steps that flow in a logical sequence from identification to formulation to implementation to evaluation. This model emphasizes a downward transmission of authorized decisions and stresses instrumental action, rational choice, and the force of a legitimate authority. "Policy then becomes a simple technical process separate from political debate, emerging as a process of technical – bureaucratic decision making guided by the political priorities of an elected government" (Scoone 2003: 1). This model assumes that policy makers approach the issues

rationally, going through each logical stage of the process, and carefully considering all relevant information. If policies do not achieve what they are intended
to achieve, blame is often not laid on the policy itself, but rather on political or
managerial failure in implementation (Juma and Clarke 1995) – failure is the
result, for example, of a lack of political will, poor management, or shortage of
resources.

Thus, the linear model is based on assumptions of rational and instrumental
behavior on behalf of decision makers (Simon 1957). The focus is on the decision and the subsequent stages of implementation (cf. Easton 1965; Jenkins 1978;
Hogwood and Gunn 1984). Such linear models offer a prescriptive, essentially
top down solution to how things should work (Sabatier 1986). These approaches
make an important distinction between processes of decision and processes of
execution, which means that this "classical" model separates policy formation
from implementation. It is operated through the linear translation of policy guidelines into implementation programs and distinguishes those who make policies
from those responsible for actually implementing them. Government policies
are usually implemented by "subordinate administrators whose obedience to
commands should be prompt, automatic, and unquestioning" (Thompson 1961:
11). Policy implementation in this case is divorced from policy and does not
form part of the policy design. Awareness of this distinction has a long history
in social science, dating at least to Max Weber's belief that as societies become
more complex and differentiated into areas of specialization, the 'iron cage' of
rationalization and bureaucratization must inevitably spread into those domains
(Weber as cited in Gerth and Mills 1991). This assumption that the organization
of all human aspects of life would become progressively smoother and more
efficient has proved problematic. Even for those writing within the public administration tradition of public policy who have some level of commitment to the
linear model, the problems are well recognized (Wilson 1993).

A significant feature of the linear model is a focus on agenda setting and on
policies as statements based on determinate decisions taken by policymakers. Yet
as Barker (1996: 27) points out, "Policy is never made once and for all. A policy
may change in the process of implementation; the intended outputs and outcomes
may not at all be those which result and those who were intended to benefit ... are
not always those who benefit." Lipsky's work on "street level bureaucrats"[1] demonstrates the great extent to which the discretion of those who implement policies
impinges on the form these policies take in practice. Similarly, Houtzanger's
(1999) research demonstrates how "policy change" may in itself arise and be
driven from "below" rather than through decisions that are made from "above."

Another critique of this classic model is that although decision making is cast within the linear model as the "bounded, purposive, calculated and sequential events which the actors perceive to have significant consequences" (Weiss 1986: 221), it is rarely possible to isolate a clear-cut group of decision makers or a particular event that can be pinpointed as the moment when the decision was made. Weiss (1986: 222) argues that people may not a priori perceive themselves as making policy: "over time, the congeries of small acts can set the direction, and the limits, of government policy. Only in retrospect do people become aware that policy was made." Rather than a decisive move toward a new agenda, policy making frequently involves marginal adjustments to existing options or simply "muddling through" (Lindblom 1959). Indeed, Clay and Schaffer (1984) go so far as to suggest that policy making lacks not only linearity, but any sense of rational, programmatic action, contending that "The whole life of policy is a chaos of purposes and accidents" (cited in Keeley and Scoones 1999: 33).

Gordon, *et al.* (1993:8) suggest that the linear model is in fact a "'dignified' myth" often shared by policy makers (see also Shore and Wright 1997). As a normative representation, a rational, technical view of the policy process masks the management of uncertainty and the politics of interactions between different agents, positions, and interests in the shaping of policy in practice.

The importance of the linear model in actual policy process literature, however, is not to be underestimated. Proponents believe for example, in the context of environmental policies, a top down, instrumentalist perspective may be appropriate for analyzing simple, easily monitored and controlled regulatory policy issues set within a well-enforced legal framework. But critics argue that when one is looking at complex, uncertain, and varied contexts of water resource management, an emphasis on local negotiation and incremental field level action, by contrast, may be more appropriate.

Although the linear scheme is useful up to a point, there is plenty of evidence that things do not work in such a tidy fashion – policy comes from many directions, and implementation can be as much about agenda setting and decision making as execution of decisions. A large and varied literature informs this thinking, ranging from structural analysis of political interests, to discursive approaches to understanding power and knowledge, to actor networks, agency and practice. Collectively the literature on understanding environmental policy processes (Keeley and Scoones 2003), policy narratives (Roe 1991), policy networks (Jordan 1990), discourse coalitions (Hajer 1995), mutual construction (Shackley and Wynne 1995), epistemic communities (Haas 1992), policy space (Grindle and Thomas 1991), and policy communities (Wilks and Wright 1987; Coleman and Skogstad, quoted in Atkinson and Coleman 1992), shares a more nuanced view of policy that sees policy making as a distinctly nonlinear process shaped over a period of time. Both in its formulation and in its implementation, the policy process involves the interaction of actors in a variety of networks. Shifting the focus away from "policy makers" to a much broader constellation of actors who engage in various ways with the process of making and shaping policy brings these dynamics into clearer view.

Actors, interests and interfaces

A focus on policies as courses of action that are part of ongoing processes of negotiation and contestation between multiple actors over time therefore provides a second approach to understanding policies (Dobuzinskis 1992). This approach focuses on the idea that "actors interact and bargain with each other and thereby produce a policy outcome" (Mooij 2003).

Viewed from this perspective, policies may not even be associated with specific decisions, and if they are, they are always multiple and overlapping. Lindblom (1959), for example, described policy making as the "science of muddling through" and advocated an incrementalist perspective on policy process (Braybrooke and Lindblom 1963; Lindblom 1979; Dror 1964; Etzioni 1967; Smith and May 1980) that focuses on the actions of policy actors and bureaucratic politics. Such a perspective suggests a more "bottom up" approach to policy (cf. Hjern and Porter 1981), emphasizing the agency of different actors across multiple "interfaces" (cf. Long and Long 1992). Long and Long's approach conveys the idea that some kind of face-to-face encounter between individuals with different interests, resources, and power brings out the dynamics of the emerging struggles and interactions that take place. Here an analysis of practitioners and their day-to-day dealings with policy issues (cf. Schon 1983) offers a key insight into the timing of "trigger events" and the role of "policy entrepreneurs" in pushing policy discussions in new directions (Cobb and Elder 1972; Kingdon 1984).

Several policy researchers who focus on interests and interaction in the formulation of policy use terms such as "policy networks," "policy communities," and "epistemic communities" (Haas 1992; Howlett and Ramesh 1998; Smith 1993a, 1993b). An approach to policy processes that puts actors into the picture helps to highlight the importance of networks through which certain versions of reality gain credibility and legitimacy and through which "policy entrepreneurs" (Kingdon 1984) are able to open up spaces and create constituencies for a particular course of action. Although these various concepts do not refer to exactly the same phenomena, Mooij (2003) argues that "they are similar in the sense that they refer to people who share ideas and are influential in setting policy agendas."

Exploring the strategies and tactics these particular kinds of actors draw on to attempt to shape the direction of policy offers insights that further undermine the linear policy narrative, particularly the causal assumption that implementation starts only after a new policy agenda has been set. As Keeley and Scoones (2000) demonstrate, closer attention to the dynamics involved would suggest that policy change emerges from the successful strategizing of networks of actors who articulate and extend alternative actions through their interpretations of reality. As such, they suggest an alternative model that practices make and shape a dynamic and malleable policy agenda rather than just respond and react to it.

Colebatch (1998: 23) suggests "a great deal of policy activity is concerned with creating and maintaining order among the diversity of participants in the policy process. It seems to be not so much about deciding, but more about negotiating ... for a common ground." Such attempts to maintain order, Colebatch observes, do not take place simply around negotiations for a common position. They also extend to attempts made by lower level workers to create some order for themselves in their jobs to gain stability and predictability within a situation in flux. This emphasis on negotiating and bargaining and on the agency of "street level bureaucrats" further disrupts the mechanistic model of a linear sequence.

Actors also resort to purposeful action to influence policy processes and to set the policy agenda (Dearing and Rogers 1996) or to transform the situation in the direction that they envision (Benford and Snow 2000). Actors enroll or enlist other actors to gain support for their interpretations (Callon 1986). Alliances that particular actors have with others in different kinds of organizations which may serve to create "room for maneuver" (Clay and Schaffer 1984) within, as well as routes for influence beyond, offer useful entry points for making sense of these complexities and processes of inclusion and exclusion, of contestation and consensus, through which particular policy positioning is shaped. Once the claims that these actors have made are supported by other actors, their claims become recognized as facts leading to a shared understanding of reality (Latour 1987). Actors forge alliances with others to mobilize larger pools of resources so that they have the power to bring about particular social, political, and economic outcomes (Long and Long 1992; Zald 1996).

This approach provides a broad framework for thinking about the type of policy communities that are able to bring about changes in policy positions but it fails to explore the social relations and micro level interactions involved. An intentional emphasis on categories of actors like the state, civil society, donors, the community, and scientific establishment involved in the policy process obscures the extent to which existing institutions condition the shaping of policy as practice. It is by examining the actions of different actors in their social relations and cultural norms and values that we are able to understand how received wisdoms get built and are subsequently upheld in policy issues. This approach is also not without criticism. Kling (1996) and Dowding (1995: 137) provide two critical reviews of the policy network processes. According to Dowding, "networks" are metaphorical, heuristic devices, which cannot explain policy processes. Their power lies in the strength and reach of the network. Power in this interpretation is instrumental and people use their power to get things done. Those who have more power are likely to win. Based on the resources that individuals, groups and networks have they can be better or worsen a situation and this influences their power to affect policy processes.

Another greatly discussed contribution to the field of policy studies has been that of the Advocacy Coalition Framework (ACF) developed by Sabatier (1987) and his associates (Sabatier and Jenkins-Smith 1988; 1993). They argue that policy studies have failed to produce a systematic body of empirical research findings. Sabatier and Jenkins-Smith (1993) argue for a causal model that offers a clear basis for hypothesis testing. ACF has its roots in policy network theory and policy communities' theory. It clearly maintains that policy making is an ongoing process with no clear demarcated beginnings or terminations. The advocacy coalition involves a broader set of processes than the policy network theory. ACF is conceived of as an alliance of political groups in a policy subsystem sharing the same interests and ideas that come together to argue against other policy coalitions concerned with the same policy issues. It involves more actors than traditional policy studies and includes journalists, public interest groups, researchers, policy analysts and state and local officials, among others. However, in contrast to policy network theory, ACF members bargain and form alliances within a policy subsystem. The goal of this theory is to explain policy change through interactions of competing advocacy coalitions.

Sabatier and Jenkins-Smith (1993: 5) refer to coalitions of "actors from a variety of … institutions at all levels of government who share a set of basic beliefs … who seek to manipulate the rules, budgets, and personnel of governmental institutions in order to achieve these goals over time." A policy position in ACF is grounded in claims about causal knowledge. The essential policy practices are resistant to change and change occurs only when large scale external shocks such as oil crises are seen to be capable of shaking the core of a policy belief system (Fischer 2003: 96). They argue that social and economic factors external to the policy making system serve as stabilizing factors for long durations but major disruptions of these conditions can alter meanings of particular policy ideas. A primary contribution of ACF research is the finding that the role of interest over time is not strong enough to explain policy change. Policy ideas and beliefs also play a major role.

However, ACF has its own limitations. The main criticism leveled against it is its inability to explain policy change. It seems better at explaining policy stability given its reliance on external events to explain political changes amongst coalitions. Overemphasis on external factors makes Sabatier neglect the role of strategies of coalition formation and the role of rhetoric and discourse in policy development. Sometimes external events fail to have effects on coalitions.

Another argument made against ACF is this framework is very US centric, most societies – even Western European countries – are more closed than the United States. While European countries are trying to change direction they are showing more fragmentation as the number of players in the policy making process increases. In developing and underdeveloped countries there are further issues of education and literacy and the process is even more closed.

ACF is also criticized for its emphasis on technical considerations advanced through expert discourses. But even in this process sometimes there exist missing empirical data which makes the empirical analysis of ACF generally infused with interpretations, despite ACF's aversion to interpretive methods. Yet ACF's contribution to policy studies cannot be denied. It brings to policy approach a new coalition of actors other than the traditional institutional groups. This approach works well in a rigorous empirical causal model of policy change, but this is not a workable solution from a constructivist perspective (Fischer 2003: 100). Policy learning in ACF is rationalistic, technocratic understanding. Hajer's critique of ACF is centered on the contention that the ACF coalitions are too thin analytically and cannot adequately account for the interactive dynamics of policy change. He argues that while ACF describes important aspects of policy change a key issue from a discourse perspective of *why* and *how* change comes about is left unexplained. It neglects social and historical factors and emphasizes empirical evidence and a desire to develop empirical hypothesis that is universally applicable to a wide range of social contexts. It considers the coalition as a relatively unified group but neglects critically important differences, from those taking a reformist perspective to those advocating a radical oppositional viewpoint (Fischer 2003: 101).

Policy as discourse

A third way of understanding policy processes is to look at policy as discourse. Discourse analysis helps to make sense of the interplay of knowledge and power in the policy process whereby particular versions of reality are adopted and upheld, and to locate the emergence and shaping of these versions in particular social and historical contexts. As Gasper and Apthorpe (1996) point out, discourse is understood and used in a range of different ways in the policy process literature.

Discourse, according to Hajer (1995:44), is "a specific ensemble of ideas, concepts, and categorizations that are produced, reproduced and transformed in a particular set of practices and through which meaning is given to physical and social realities." These "ideas, concepts, and categorizations" constitute expressions of knowledge and power (Foucault 1980). Discourses are thus frames that define the world in a particular way and which exclude alternative interpretations (Schram 1993; Grillo 1997). Discourses do not arise from particular individuals but are the cumulative effects of many practices.

In relating discourse to policy, two approaches have particular salience in making sense of all these processes: the first is derived from the analysis of texts and utterances and involves the deconstruction of terms used in policy language. The way issues are talked about is highly important. The key assumption behind this approach is that "rival ways of naming and framing set policy agendas differently" (Gasper and Apthorpe 1996: 24), hence a closer look at the terms and concepts and at the stylistic devices that are deployed in framing the objects and scope of policy offers a productive entry point for understanding how particular interpretations of reality come to gain hegemony. Concepts also have histories, which reflect types of knowledge that empower certain institutions and individuals and simultaneously marginalize others (Dryzek 1997). They also guide policy in certain directions.

Gasper and Apthorpe (1996) highlight a range of rhetorical and conceptual devices in "policy-speak": good/bad binaries to mark out normative positions, metaphors and allusions, nouns rather than verbs, normative rather than descriptive terms or key words and slogans to bolster "grand strategies." Numbering and naming serve to bound and defend particular versions. By highlighting the style, form, and language used in the construction of policy statements and in the interactions that shape policy processes, strategies such as deconstruction and narrative analysis complement the notion of discourse and prove valuable in policy analysis. Set within the social and historical context in which ideas are generated and stories are created and told, discourse analysis can provide insights into how particular stories gain ascendancy and others fall by the wayside.

A second approach to the analysis of policy discourses has a wider purview. Many analysts turn to what has been termed the "argumentative turn" in policy analysis to draw attention to the ways in which particular concepts or storylines "frame" what and who is taken into consideration in, and excluded from, policy deliberations (Fischer and Forster 1993; Hajer 1995; Rein and Schon 1993). The framing approach extends from semiotic or narrative analysis of policies themselves to a discourse analysis of the role of different actors in the policy process. It is here that approaches to discourse informed by the work of Michel

Foucault (1977, 1979) become particularly valuable in making sense of policy processes.

The Foucauldian concept of discourse refers to a historically situated set of practices that produce and reproduce relations of knowledge and power. Power for Foucault is imminent in all social relationships. For Shore and Wright (1997), analyzing policy as discourse is perhaps most potent in drawing attentions to the ways in which the political nature of policy making is camouflaged by recourse to idioms of objectivity, neutrality, and rationality. Drawing on Foucault's notion of "political technology,"[2] they argue against the instrumental view of policy, which treats it as an instrument of governance, a rational, non-theoretical and goal oriented tool that provides the most efficient means to obtaining certain desired ends (Shore and Wright 1997: 28). According to Shore and Wright (1997: 29) "policy" is always informed by ideological considerations and often codifies morality by functioning as a Foucauldian "political technology" that masks its political origins and the relations of power it helps to reproduce. These political technologies, according to Dreyfus and Rainbow (1982: 196), "advance by taking what is essentially a political problem, removing it from the realm of political discourse, and recasting it in the neutral language of science" (cited by Shore and Wright 1997: 37). This principle has important implications for the ways in which information and knowledge come to be represented in the policy process.

In the most comprehensive articulation of different approaches to policy, one that draws on Foucault's conception of discourse to analyze the manner in which policies are constructed and extended, Hajer contributes two key concepts: those of "storylines" and "discourse coalitions." Storylines, according to Hajer, gain their discursive power by combining elements from different domains to provide actors with a series of symbolic references that suggest not only a common understanding, but also one that sounds "right." This sense of "right-ness" is influenced as much by the trust people have in the "story teller" as in the persuasiveness of the story itself, and the acceptability of the story for their own identities. Storylines, then, not only cluster knowledge but position actors, proving "the essential discursive cement that creates communicative networks among actors with different or at best overlapping perceptions and understand-ings" (Hajer 1995: 61–3).

Hajer accounts for how particular storylines are taken up in policy through an analysis of "discourse coalitions," through which previously independent practices are actively brought together and given meaning within a common political project. Through these coalitions, actors "not only try to make others see the problems according to their views, but also seek to position actors in a specific way" (Hajer 1995: 53). This way, according to Hajer, a picture emerges of competing discourses being shaped into a single narrative, with political contests taking place around the framing and promotion of particular storylines. Discourses and framing devices used in policy deliberations highlight the ways in which particular versions of reality gain legitimacy. Where Hajer differs from Sabatier in discourse versus advocacy coalitions is his emphasis on interpretation versus empirical cognitive dimensions. For Sabatier and Jenkins-Smith (1993: 5) ACF involves individuals who get together to coordinate political activities in specific, identifiable ways based on a technocratic vision and quantifiable indicators. But Hajer believes that discourse coalitions are a discursive phenomenon and are "reproduced and transformed through a variety of political actors that do not necessarily meet but through their activities reinforce a particular set of storylines in a given policy domain" (Fischer 2003: 105). Fischer (2003) argues that people in India, Germany and the United States, for example, can all share, sustain, and reproduce the sustainable development storyline without having met each other let alone having coordinated their political activities. Another example he gives is of people contributing to Sierra Club or Friends of the Earth, without ever having gone to a meeting or knowing the people in the organization. The ACF researchers emphasize the positive benefits of quantitative indicators in helping solve policy controversies; discourse theorists investigate why these particular indicators are seen as legitimate and appropriate for the role. Discourse analysts have to dig out different ways of talking that can be found in the environmental domain rather than an objective environmental problem that can be nailed down empirically in ACF. Instead of organizing research to facilitate the search for empirical generalizations, the discourse analyst looks at a detailed contextual examination of the circumstances at play in specific cases as a key to explaining how change cones about. Thus, the making and shaping of policy becomes a "political technology," relying on the versions of "expertise" and institutional techniques that create and define the category of reality.

For the discourse literature, power lies not in the individual but in the discourse itself. Discourse is the sum of numerous micro practices, and it is through these practices that power is exercised. It works through interpretations and meanings. Together these practices form a powerful discourse that frames how people act and think about the world. It controls processes and frames reality in such a way as to establish conditions that determine the possibility for change. An element that adds to the stability of these discourse coalitions is the role of credibility, acceptability and trust – instead of just empirical evidence in explanation of policy change. While ACF claims that science can bring about a consensus on policy matters based on quantitative data and comparison of objective findings, the social constructivist discourse analysis points to the large amount of research that shows that such findings are intermingled with discourses about political acceptability and social trust, especially in policy forums (Fischer 2003: 133).[3] A discourse perspective on policy process, Keeley and Scoones (1999: 32) believe, might suggest that options for participatory forms of policy process are highly constrained, as some would argue that deliberative policy processes are just extensions of state discourse and power under the guise of participatory rhetoric. It would also override potential for debate and participatory deliberations on policy issues due to its strong political, business, and bureaucratic structure. However, they believe that such a universalized view of discourse conceals other dynamics. There is no doubt that in relation to any particular policy issue, competing and multiple discourses exist based on their localities and coalitions of people under whom, as Hajer (1995) argues, opportunities for an argumentative turn in policy analysis can open up. In this manner, spaces are created in the policy arena for new actors and voices. In political theory agency is becoming important as structural attempts to explain political interactions look increasingly tired.

Policy spaces

Grindle and Thomas's (1991) widely cited concept of "policy space" assumes that since policy making remains an essentially indeterminate and unpredictable process of engagement in which outcomes are always uncertain and contingent, the possibilities for change and alternatives to emerge cannot be bounded off. Policy space then emerges in the form of events, and actors challenging pervasive orthodoxies reframe the debate and reconfigure relationships between actors. This process becomes an important entry point to understanding the dynamics of alternative narratives and agendas of particular reforms in policy.

Literature on policy process, however, reveals that the growth of this branch of policy studies is a relatively recent phenomenon, taking place over approximately the last 15 years. Most of the literature on policy processes is dominated by examples from the United States, and to some extent Great Britain, rather than from developing countries (Mooij 2003). One analysis that specifically addresses the question of a third world policy process is a paper by Horowitz from 1989 (Mooij 2003: 6) that mentions that the policy process has some commonalities and contrasts in developed and developing countries. There are also important differences in the policy process that need careful consideration while making broad generalizations (Horowitz 1989). What is noteworthy according to Mooij and Vos (2003) is that the study of policy processes in developing countries is in part stimulated by international donors and research institutions closely linked to development aid agencies and donors that provide funds for their projects. For example in Britain, the Institute of Development Studies (IDS, Sussex) and the Overseas Development Institute (ODI, London) produce interesting work on policy processes in developing countries, but most of it is funded by the Department for International Development.

Policy process literature in India

In India this branch of the literature has started to receive attention since the 1980s. In the discipline of political science there exists a sheer neglect of the systematic study of public policy as political scientists have been engrossed in the study of political institutions and processes (Arora 2002: 46). While much has been written about policy making in India, relatively little has focused on the emergent dynamics of policy process in India in general and more specifically in the post-economic reform era (Scoones 2003a: 2, 2003b). In India, there is a considerable amount of literature on political policy processes, but it does not deal with policy per se. Works on general aspects of the politics of policy and evolution of policy process in India include Harriss (1988), Arora (1993), and Mathur (2001). Harriss (1988) in her book focuses on food policy implementation in India and the insecurity that arises from policy implementation. Arora's (1993) paper of the policy regime in India is directed towards the post-economic period of 1991 and the corruption and victimization that occur in the new policy regime. Mathur (2001) on the other hand traces the evolution of public policy in India. These processes, Mathur argues, are constrained by the fact that India still is in the process of moving towards a more participative democracy, despite having democratic institutions since independence. This affects the policy process in two different ways; politicians are more concerned about their political survival than policy, while the bureaucracy is massive, self-interested and not very effective. According to Mathur, these political and bureaucratic characteristics have made the Indian policy making process a technocratic venture. Policy formation and choices accordingly emerge from specially appointed committees, which consist of the political and civil service elite who seek a quick policy recommendation. The rise of nongovernmental organizations (NGOs) that have started to challenge policy and offer alternatives is a recent phenomenon but their role is constrained by political and administrative concerns. Mathur's article provides a general overview of policy making process in India without any specific reference to case studies and is also silent on the interactive dynamics of policy processes. Beyond these general perspectives most of the work that can be related to policy processes in some way involves the economic reform era of India that began in 1991.

Literature on the politics of the economic reform process

Most of the literature available on aspects of policy process in India is on the politics of the economic reform process in the country. Some of the most prominent analyses on different aspects of the liberalization process and the politics of the reform process and its acceptance in the system include Harriss (1985), Kohli (1989), Kochanek (1996), Shastri (2001), Weiner (1999), Bhaduri and Nayyar (1996), Jenkins (1999) and Patnaik (2000). A brief overview of their arguments will help explain the limited nature of research on policy per se as well as identify unexplored policy processes. Harriss (1985) argues that India's efforts to liberalize in the 1980s failed due to the compromised nature of class power. There existed an uneasy alliance between India's bourgeoisie and its dominant rural classes, which led to a crisis in planned development. This process constrained the state to move ahead and change the course of India's development policy. Kohli's (1989) work is an analysis of why liberalization was adopted but the focus is particularly on domestic policy processes, and external variables remain outside the scope of the paper. Shastri's (2001) paper makes a relevant point in investigating the sources of policy change in India in the 1990s in the liberalization process. The author's emphasis is on ideas as a source of policy change. Shastri makes the point that the shift in the policy discourse on liberalization was pushed by a changed team that consisted of key decision makers and economic advisors having a liberal ideological orientation based on their connections with the World Bank, the IMF and Western institutions.

Weiner's (1999) article analyses the role of the individual state Governments in India's economic liberalization and the way in which federalism and centre–state relations helped to accelerate or slow down the reform process. The focus is more on center–state relations and the centralizing features of the quasi-federal system in India. It is due to this centralizing federalism that reforms have moved slowly in the states. However, Weiner argues that given the nature of the political and environmental situations, states must compete with each other not only in New Delhi but also in the global market place. This, according to Weiner, can only be accomplished by a new political mindset and a more efficient state bureaucracy. Bhaduri and Nayyar's (1996) book is a critique of the economic reform process in India. Another article that critiques the neoliberal policy on many fronts, but also addresses the question as to why these policies have become accepted in India, is the work of Patnaik (2000). He makes the point that independent academic work is difficult to undertake in an increasingly market driven world which leads to an abridgment of democracy. Policy making thus takes place in a strategic way, through which public opinion is shaped. The works of Jenkins (1999), Corbridge and Harriss (2000) and Kohli (2001) assess new patterns of politics and governance to some extent in the post-reform era but most of these works remain silent on the key interactions of various actors, knowledge, and power in the new economic era. Classic treatments of policy analysis literature can be found in Frankel (1978) and Rudolph and Rudolph (1987), but while these sources offer useful insights into the study of policy, they do not look extensively at policy processes. Other works on governance in Andhra Pradesh, (Sudan 2000), state policy making on food and the politics of hunger (Currie 2000; Mooij 1999a, 1999b), politics in the policy process in Andhra Pradesh (Mooij 2003), the state and the poor (Echeverri-Gent 1993; Kohli 1987; Manor 1993), drought (Mathur and Jayal 1993), and agricultural policy (Varshney 1989; Rao 1999, 2001)[4] can be cited as examples of state policy making in developmental areas.

A review of literature on policy process in India reveals that there some case studies but not a rich body of literature (Mooij 2003).[5] Recently attention has begun to be paid to policy making and there have been efforts to stimulate research on policy processes but this is largely funded by external agencies (ibid., v).[6] Mooij and Vos's (2003: 5) bibliography of policy process literature assesses the Indian situation as follows:

> Within India, the study of policy processes is not very well developed. This is so, despite the fact that many Indian social scientists are involved in policy relevant research and aim to contribute, through debate and research, to policy formulation and implementation. These debates are, however, almost entirely dominated by economists, and insights from other social sciences have hardly entered into them. There are very few political scientists, sociologists or anthropologists focusing on public policies. As a result, some aspects of policy studies are relatively well developed (such as measuring policy effects), but others much less. The issues and questions, for instance, of why policies are formulated and designed in particular ways in the first place, and the political shaping of policies 'on the ground', do not receive much attention.

Literature on water policy

Mollinga (2005) places the literature on water into three categories: (1) literature on desirable water policy, (2) empirical research on water resource management, and (3) critical, oppositional discourse. All of these categories deal with research in policy and research for policy but, as Mollinga points out, there has been very limited research on policy processes. In the politics of water, the study of hydropolitics is, in general, one of the more researched areas. Literature focuses on the hydropolitics of the region (Iyer 2000, 2003; Upreti 1993; Gyawali and Dixit 1999; Gyawali 2000; Crow 1995; and others); on the politics of interstate Himalayan rivers in Nepal, India, and Bangladesh; and hydropolitics in third world (Elhance 1999). This work has mainly been on the interstate level.

Most of the literature on desirable water policy takes the form of policy documents, government reports, committee documents, and the like, either advocating better policy or the need to implement the existing policy in a better way. The reports of the Expert Group on Commercialization of Infrastructure Projects (EGCIP 1996), the Standing Committee on Urban and Rural Development (2002) and national and state water policy documents and texts represent the visions of other actors that network for a better policy. Nongovernmental actors can also produce documents for a desirable water policy. For example, some literature on desirable water policy can be found in the Centre for Science and Environment, New Delhi, a NGO that works in the area of the environment. Publications from the Centre argue for making water everybody's business, advocating a policy of water harvesting using traditional knowledge systems, and encourage community management of water resources (Agarwal, Narain and Khurana 2001). Think tanks like the Institute of Policy Research, New Delhi, suggest new alternatives for the future of water usage (Iyer 2000, 2003, 2008). Another category is the loan and project documents that stipulate what institutional changes will be implemented as part of a certain set of activities. An example is the institutional conditionalities of the Asian Development Bank and the World Bank supported water projects in different states. Together these documents and their presentation and discussion in the press, parliament, committees, workshops, and elsewhere could be said to constitute the "formal," that is, government connected, water policy discourse (Mollinga 2005: 9).

The second category identified by Mollinga (2005), empirical research on water resource management, includes a relatively rich literature on actual water use and management at the field/user/consumer level, and its impacts. One set of literature in this section includes the (agricultural) economics literature looking at institutions and incentives that shape water resources management, but explicit discussion of the relations of social power is still rare in economics oriented papers. Literature in this area also includes a lot of work on water resource management in India that has been done by the scientists, economists and engineers in the International Water Management Institute, which focuses on irrigation management, groundwater depletion issues, and participatory research in this area.[7] Other examples are Mollinga, *et al.*'s (2001) work on the implementation of participatory irrigation management in Andhra Pradesh; Vaidyanathan's (1999) work on water resource management institutions and irrigation development in Indian tank irrigation systems (Sridhar, *et al.* 2006); a top down to bottom up study on institutional reforms in Indian canal irrigation (Gulati, *et al.* 1999); and irrigation policy and practice in India by Maloney and Raju (1994). Zerah's work (2003, 2000a, 2000b) on urban water looks at urban water scenarios in Indian cities and its impact on households and at informal and formal water suppliers in cities and their impacts The other literature looks at the social relations of water resources management, often taking on the issue of unequal social water management practices in India, including dam conflicts, social movements, and mobilization.

Mollinga's (2005) third category of research is on critical oppositional discourse. This literature points out the problems and failures of the present water resources policies and institutions, surveying environmental critiques, gender critiques, privatization critiques and others, and calls for a paradigm shift in the sector, often in the context of a broader "alternative development" perspective. Much of this literature has emerged from the grassroots struggles for the management of common property resources, which has been an important yet complicated issue in the management of water resources. Sharma (1998) provides an anti-state perspective to resource management issues while trying to validate traditional forms of water harvesting and its control through grassroots actors. Much work exists on opposition to dams, which forms part of this oppositional discourse. Ali-Bogaert's (1997) work emphasizes the development discourse in India and a counter-discourse of resistance to the development of the Narmada Bachao Andolan NGO against the Narmada Dam projects. Khagram (2004) and McCully (2001) provide other examples of work on the politics of large dams and the transnational struggles surrounding them.

However, in accordance with Mooij's view that policy process research is donor driven, Mollinga (2005: 10) also feels inclined to argue that the more innovative research work on water resources policy and management increasingly tends to come from "'nonconventional sources,' that is, not from universities and established research institutes." Mollinga's (2005) and Mollinga and Boulding's (2004) work calls for a research agenda on water policy process in agriculture and irrigation management. Research by the IDS[8] in which Lyla Mehta[9] and others have examined various positions on water scarcity in Gujarat – including gender and forced displacement, rights of the displaced, access to natural resources and power/knowledge interfaces in policy debates in the Sardar Sarovar Dam project – does provide inputs in this policy area but in general, there is a dearth of literature available on water policy processes in India.

The literature reviewed suggests that there is limited work on policy process in

India in general and water in particular. The Delhi water reform project provides the right lens for conceptualizing policy as a process in the post-reform era. To understand, therefore, the ways in which water knowledge affects water policies one must therefore analyze the policy process as one of multiple spaces of contestation involving a complex configuration of actors, discourses, and knowledge. Rather than denying the existence of these different interests and perspectives by adopting a classical model, an attempt to capture these different interests and perspectives in the politics of policy making and the nature of power in policy making can be achieved.

This approach to policy has several advantages: (1) it goes beyond the narrow, legalistic, formal and written aspects of policy; (2) it is not prescriptive, but subjective and perceptual; (3) it captures different interests and perspectives of actors; (4) it is able to capture the incremental and complex nature of policy development; and (5) it taps into policy discourses and advocacy coalitions concerning the legitimacy, realism and efficiency of policies (Pasteur 2001). Figure 2.1 illustrates the framework of analysis which helps to make sense of the interplay of knowledge and power in the policy process where particular versions of reality are formed and maintained. It also locates the emergence and shaping of these versions in particular social and historical contexts.

This framework not only includes those voices and views involved in the policy process but also those that are left out of the formal policy processes, because these actors also make claims to valid knowledge, around which they organize themselves. Actors at varying levels use differing discourses and water knowledge to mobilize support for their claims. Understanding how power, knowledge, and agency define spaces for engagement, privileging certain voices and versions and excluding others in water policy processes comes from examining how differing discourses and actors interact in such spaces. This framework of competing discourses provides an understanding of how particular ways of thinking about water have gained ascendancy and determined the frame through which water is defined, measured, and approached in Delhi's urban water reform policy.

The sheer complexity of the web of actors engaged in policy processes, whose

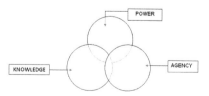

Figure 2.1 Conceptual framework.

connections and interactions weave across and within the artificial divide between "citizens" and the "state" make the process of understanding the nature of policy change extremely complex. Different actors within and outside the state and civil society may take up a range of subject positions and represent a constellation of competing interests to represent their knowledge claims. These actors both use and are bounded by variant discourses of reforms in urban water policy. By examining how different narratives of water and different actors interact in such spaces – as well as how they may be included or excluded from them – this framework allows a better understanding of the ways in which power, knowledge, and agency frame policy processes. Using this framework we now turn to the economic reform era to trace the policy processes of economic liberalization and private sector participation in India.

Notes

1 Lipsky (1980) refers to the role actors who implement policy changes have to play in the process. He emphasizes that they have substantial ability to mold policy outcomes. Street level bureaucracies are schools, welfare departments, lower courts, legal service offices, etc. As a result of time constraints and other practical considerations, as well as political opinion, those who work in these bureaucracies influence the practical working out of a policy to produce an outcome that may be substantially different from that originally intended by a policy maker.

2 First introduced by Foucault, this term relates to the way policy is often "depoliticized" if such depoliticization is in the interest of the dominant group. A political problem is removed from the realm of political discourse and recast in the neutral language of science. It is represented as objective, neutral, value-free, and often cast in legal or scientific terminology to emphasize these characteristics. This depoliticization reflects the "technology of politics" by which various means are used to work within a political agenda. This masking of the political under the cloak of neutrality is a key feature of modern power (Shore and Wright 1997).

3 In *Swimming Upstream: Collaborative Approaches to Watershed Management* (Sabatier, *et al.* 2005), Sabatier does explore issues of trust and legitimacy in relation to stakeholders' trust of other stakeholders as well as stakeholders' trust of policy officials. A watershed management policy, he argues, takes repeated interactions over time to build trust which is important to increase the efficiency and effectiveness of the government and to implement policy outcomes without coercion.

4 For a detailed bibliography on policy process in general and policy process in India one can refer to "Policy processes: An annotated bibliography on policy processes, with particular emphasis on India. Working Paper 221" by Jos Mooij and Veronica de Vos (2003), published by the Overseas Development Institute, London, UK.

5 Most of the work is done by the Institute of Development Studies in Sussex, UK, a lot of which is donor funded to stimulate work in developing countries like India.

6 These centers are the Centre for Law and Governance (part of the Jawaharlal Nehru University), the Centre for Public Policy and Governance (related to the Delhi based Institute of Applied Manpower Research) and the Centre for Public Policy (part of the Indian Institute of Management, Bangalore). The first two have received money from the Ford Foundation; the last from the UNDP (Mooij, 2003).

7 The websites of the International Water Management Institute (http://www.iwmi.org)

and the Comprehensive Assessment of Water Management in Agriculture (http://www.iwmi.cgiar.org/assessment) provide detailed information.

8 As mentioned earlier, a lot of research is funded by the UK Department for International Development and external agencies.

9 Lyla Mehta's work has focused on policy processes and the social construction of water as a scarcity to justify developmental projects like the Sardar Sarovar Dam which is based on a technocratic supply based paradigm.

3   The process of economic liberalization and private sector participation

Making sense of policy processes requires an understanding of how power and knowledge define spaces of engagement – privileging a few and excluding the others. This task calls for a historical perspective that situates contemporary water policy reform in India within the larger processes of neoliberalism, an ideology of market forces and decentralization that has led the Indian state to turn away from the Nehru–Mahanalobian[1] socialist model of centralized development planning and adopt the politics of globalization and economic liberalization. This chapter will examine how these new ideas have gained ascendancy in the area of policy making and profoundly influenced the way water is perceived, defined, and managed.

Since water in India is under the authority of the individual states, this chapter sets out the context in which the state Governments' water policies shifted toward private sector participation, which is an integral element in the neoliberal ideology that has come to dominate the way decision makers think about government and governance in India. The task here is to understand how political interests, policy entrepreneurs, and external factors led the states in India to retreat from the vast body of regulatory controls and public management policies that had prevailed for decades, and move toward a fundamentally different way of perceiving the role government should play in the critical task of providing water to the population.

In this book the origins of economic liberalization in India are traced to the 1980s during the regime of Prime Minister Rajiv Gandhi – as opposed to the general assumption that it all began with the fiscal crisis in 1991. This chapter elaborates on the political leadership – its ideas, its key people, and its ideological relationships – focusing in particular on the alliance that was formed among a like-minded bureaucratic elite, known as the "Change Team," which gave a new direction to the economic policy making process. It further discusses the context and history of critical events during which the liberalization program replaced India's socialist developmental policy basis in a manner that had to be "politically managed" and "socially accepted" to carry out the tasks that would bring about a "shining" India. Economic liberalization in India was based on what Nayar (2001) calls an "ideas and power model."

The process of economic liberalization

After centuries of colonial rule and expropriation of resources, national economic objectives in the era immediately following colonization emphasized the nationalization of extractive and core industrial sectors of the Indian economy as a strategy for economic development. The first Prime Minister, Jawaharlal Nehru, opted for a policy based on centralized planning and development. He envisaged a capital intensive strategy to achieve both growth and social justice in the country. However, India's economic performance was "starkly unimpressive in terms of per capita income, alleviation of poverty, and share in the world output and export" (Nayar 2001) from independence until the demise of the Nehru model in the 1980s, when the Congress Party Government, which had strongly championed Nehru's socialist ideology, performed an about-face and adopted a policy of economic liberalization.

The process of liberalization began under Prime Minister Indira Gandhi, the leader of the Congress Party from 1980 to 1984. When Indira Gandhi came to power in 1980, the Indian economy was in urgent need of new economic policy initiatives. A severe drought caused a 15 percent drop in agricultural production and a decline in hydropower production leading to severe energy shortages during 1979–80. Oil supply was disrupted due to agitation in the oil-producing state of Assam. A second oil crisis during the same year increased the country's import bill, contributing to a 20 percent inflation rate and a substantial decline in foreign exchange reserves. Indira Gandhi made two important decisions in response to this economic situation: first she decided to pursue a policy of selective liberalization and second she obtained a loan from the IMF (Dash 1999: 891). Various committees were set up to review policies on selective liberalization. Headed by a group of senior bureaucrats whose ideological outlook had already undergone substantial change in the direction of liberalization, these committees produced the recommendations for reform in industrial policy.

While Indira Gandhi was aware that India did not face an immediate balance of payments crisis, she knew that India's growing import bill, coupled with declining exports, would lead to a balance of payments deficit in the near future. Her decision to move toward economic liberalization and to apply for an IMF loan of 5 billion US dollars was strongly opposed by the Communist Party of India. The nationalist Bhartiya Janata Party (BJP) also condemned the Prime Minister's decision as a betrayal of the nationalist cause by giving up India's longstanding goal of self-reliance. However, she was able to "placate the opposition party because of her comfortable majority in Lok Sabha [the lower house of the Indian parliament] and demonstration of unity by its members" (Dash 1999: 892).

Since the mid-1970s, most businesses had been demanding liberalization measures. By adopting business friendly economic policies and appointing L. K. Jha, who enjoyed a pro-business reputation, as her chief economic advisor, Indira Gandhi generated significant support within the business community (Dash 1999: 893). Despite criticism from a few Leftist intellectuals, the Congress Party Government did not draw adverse reactions from professional groups and the middle classes. To maintain her socialist image, Gandhi appointed S. Chakrabarty and K. N. Raj to the economic advisory council. However, this was just a cosmetic move. Behind the scenes, pro-reform bureaucrats such as K. C. Alexander, K. N. Jha, and Arjun Sengupta made the real economic decisions (Kohli 1991: 313–15).

With these bureaucrats working behind the scenes, and business and middle class supporting liberalization policies, India applied for a loan to the IMF, which was granted in 1981. After the first two disbursements, what some economists call an economic miracle occurred and India cancelled the third installment of its IMF loan, amounting to $1.1 billion, mainly due to a good monsoon season and increased exports that enhanced the foreign exchange earnings.The entire loan was repaid in due course.

The assassination of Indira Gandhi in 1984 and Rajiv Gandhi's enormous electoral victory in 1984[2] created a space for change in policy that can be best explained in terms of state (i.e. central Government) autonomy. "The state suddenly stood quite autonomous, seemingly free of societal constraints, ready to be used as a tool for imposing economic rationality upon society" (Kohli 1991: 312). Rajiv Gandhi's landslide victory freed him from internal pressures and allowed him to use this newfound freedom to push for a radical change in development policy.

Rajiv Gandhi and economic liberalization

Rajiv Gandhi belonged to a new generation of politicians educated in and exposed to Western economic ideas. His views attracted a new generation of political aides who had managerial and technocratic experience in the private sector just as India faced a brewing economic crisis. While India had repaid her IMF loan in 1984, in the post-1985 period India failed to control the expansion of public sector employment, subsidies, military expenditures, and interest payments (Dash 1999). As a result, the Government began to borrow heavily from both commercial sources and international financial institutions like the World Bank and the IMF in order to finance its growing budget deficits. Inflation was therefore at an all-time high during Rajiv Gandhi's time, contributing to a crisis mentality.

Under these political conditions, "[Gandhi] was struggling to get away from the old style activist, interventionist state in which, of course he had taken power" (Arora 1998: 200). The need to deal with the realities of a limited budget created the conditions "to stress a new beginning," emphasizing change rather than continuity with the past (Kohli 1989: 313).

Many of Rajiv Gandhi's ideas can be found in the Congress Party's election manifesto (1991), the economic side of which was entirely drafted by him. Liberal in his approach to modernization, Gandhi critiqued the various elements of earlier consensus on economic policy and its relevance in a changing global economy. Convinced that important changes were needed in the economic strategy, he was "actually seeking a role for the primacy of the market forces" (Arora 1998: 197). His vision was to take India into the twenty-first century on par with the rest of the world. "If India is not to lag behind other countries then our basic thinking has to change" (Gandhi 1986: 65).

Another feature of Gandhi's thinking was "a complete absence of prejudice against the private sector" (Ahluwalia 1998: 203). He believed that India had "too many controls and we have lost control" (Ahluwalia 1998: 203). Aware that India had evolved a nonfunctioning system, he geared much of the thrust of reform and liberalization to remove India's excessive regulatory apparatus which had hitherto deadened any effective entrepreneurial movement against controls on India's economic development (Ahluwalia 1998: 203).

At the same time, he was highly critical of the performance of public sector enterprises: "It is of utmost importance that we rethink our systems of management in the public sector to improve its efficiency and financial viability" (Gandhi 1988: 142–3). The signals, in the words of Pranab Mukherjee (1998: 207), were that "we cannot simply support inefficiency providing budgetary support to the public sector enterprises unless at some point of the time we tell them that you have to stand on your own legs or you have to face the consequences."

Freed by his great electoral mandate from the societal constraints of the past, Rajiv Gandhi's administration became something akin to an elected dictatorship in which the visible representative of a non-corrupt, reform-minded Indian generation focused on applying modern managerial techniques. While Rajiv Gandhi's huge majority and support in the corridors of power permitted him to institute bold and decisive reforms, voices of dissent from within and outside the Congress Party occasionally sidetracked the process, and in some cases even stalled it. In the years immediately following Rajiv Gandhi's rise to power in 1985, there was little opposition to the policy shift in the ranks of the Congress Party, but in the later years of his tenure many groups started to protest. The fear of losing electoral support forced the Government to slow down the pace of economic change, and the state lost its temporary autonomy (Manor 1987). However, ideas, once introduced, have a power of their own. They find the conduits through which they can flow and gain a momentum that lasts longer than any one administration (Shastri 2001; Nayar 2001).

Ideas and interests shaping policy reform

Although political opposition prevented a complete reform of policy from taking hold during Rajiv Gandhi's regime, by questioning some of the earlier assumptions of economic development, Gandhi initiated a process of critical thinking that led to an influential – even transformative – debate on development policy (Shastri 2001: 3). Under Gandhi's sponsorship, a team of policy advisors from inside and outside the bureaucracy were beginning to agree that economic policy needed to be changed. Shastri's work (2001) speaks extensively to the idea that a number of senior officials were gradually convinced about the usefulness of market friendly ideas and therefore of the need to open up the economy. More importantly, a section of the elite came to realize that things had not progressed the way that had been visualized in the early five-year-plan documents of the socialist past.

Key constituents of this debate were the so-called "lateral agents" in the economic process. These individuals formed a group of advisors to the Prime Minister who were mainly appointed from outside the bureaucracy – from such places as the World Bank and the IMF. These economic advisors brought with them a set of ideas from their own experience and from the predominant thinking that pervaded the networks in which they operated. The change-conducive political environment in the mid-1980s provided the policy space for these "laterals" to come in and recommend that political incumbents adopt their ideas about different policy options. This new discourse eventually developed an agenda for liberalization, which in spite of protests, postponements, and differences with opponents, remained in the pipeline to be implemented. These policy advisors – also known as the "Change Team" (Waterbury 1990:191) and "policy entrepreneurs" (Keeley and Scoones 1999: 21) – forged the link between the adoption of these new ideas and interests and the policy process through which they could then be implemented. This link between theory and practice was a major factor in the continued push to liberalization after 1990. To understand how the process of liberalization actually began in the 1990s, it is therefore essential to understand how the members of this pivotal Change Team were able to see policy spaces opening up and respond to "trigger" or "focusing events" when they arose (Cobb and Elder 1972; Kingdon 1984).

The role of policy entrepreneurs

The Change Team in India, as illustrated in Figure 3.1, comprised both political and bureaucratic members, most of whom had either been educated in the West or had gone on deputation posting and trainings where they were exposed to Western ideas of liberalism and the market process (Shastri 2001). The politicians have to face the electorate and generally base their policy announcements on expert opinions on controversial issues but the bureaucrats, inducted from the Indian Administrative Service (a legacy of the Indian Civil Service during British rule) were not concerned with seeking popular mandates. Separate from the permanent civil service representatives was a key category of the bureaucratic elite known as "laterals" that were recruited from outside the services. Because the bureaucratic members of the Change Team are politically insulated and not directly account-able to the electorate, they were in a position to push policy reforms without direct confrontation with the electorate.

Both the career and lateral bureaucratic elite played a crucial role in the early phase of the liberalization process. From their protected position, they were able to continue their work from Rajiv Gandhi's administration, writing policy papers and committee reports detailing the sequence of the liberalization process while

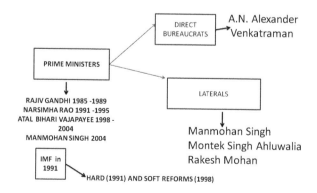

Figure 3.1 Change team: ideas and power model (1985–2004).

political actors came and went in the aftermath of Rajiv Gandhi's assassination in 1991.

Prominent among the bureaucrats of the civil services who played a major role in the reform process were A. N. Verma, Dr Abid Hussain, Gopi Arora, S. Venkataramanan, S. Ganeshan, and others. The foreign experiences and exposures of these bureaucrats in top positions in the West enabled them to critically compare India to other nations rather than comparing present day achievements with what India had achieved since the 1950s. On the basis of their experiences abroad, they were able to apply a global perspective to policy recommendations and to the framing of new policies arising out of a critical exploration of the process of economic development (Shastri 2001). As a result, a consensus began to emerge within the bureaucracy that a liberalization agenda was imperative.

Rajiv Gandhi's tenure created the political atmosphere necessary to translate ideas into action. To some he was a child of his times, but to others he was, to a considerable extent, ahead of his times. It was only because neither the political party that he presided over nor the political conditions that prevailed during his times were prepared to accept the changes in his time (Kohli 1989) that his vision was not completely realized. Although Gandhi had to backtrack on his reform policies, the ideas and convictions within the bureaucracy regarding the necessity of reform continued to gather strength (Shastri 2001). The political atmosphere created by him helped the policy entrepreneurs to sequence the liberalization process when Narsimha Rao took over in 1991. From the first generation of reforms to windows of opportunity in the second generation of reforms, the lateral entrants had a significant role to play.

The laterals were primarily educated in the United Kingdom or the United States, had diverse career backgrounds, and maintained international networks. In spite of acting in an advisory capacity in India, they were and are constantly in touch with these international networks and institutions from which the discourses of liberalization evolved. In between their stints as advisors to the Government, they are constantly attending workshops, in training, and/or occupying positions in networks of neoliberal organizations like the World Bank, the IMF, the World Trade Organization (WTO), and Washington, DC think tanks.

It is through these networks that knowledge of external influences is brought to bear on the policy making elite (Haggard and Webb 1993). These types of networks mediate policy coordination and development. The networks of support that they build are mostly informal and nonhierarchical. They are often, as Shastri (2001: 6) argues, "not ideological" and subscribe to a belief in "less government" – government that is decentralized, knowledge based, rule based, less discretionary, and less powerful in the long run.

Most of these advisors have retained a key economic place in the ministries and organizations since the 1980s, thereby maintaining continuity in the reform process. In spite of the fact that Prime Ministers and Governments have changed many times, the key bureaucratic setup has remained the same. Even the BJP Government with its slogans of "Swadeshi" ("self-sufficiency") and "Economic Nationalism" retained some of these key advisors.

While there were many advisors to the reform process, two prominent figures, Montek Singh Ahluwalia and Dr Rakesh Mohan, are especially important in the context of economic liberalization processes and infrastructure development. Both these figures have held key positions in the economic ministries, the Planning Commission, and the Prime Minister's office and have chaired several committees since 1979. In the Congress party coalition government (2004–2009) led by Prime Minister Man Mohan Singh, Montek Singh Ahluwalia and Rakesh Mohan were appointed to the positions of Deputy Chairman to the Planning Commission of India and Deputy Governor of the Reserve Bank of India (RBI).

Montek Singh Ahluwalia, a graduate of Oxford University with experience in international financial institutions, has been advisor to the Indian Government in one capacity or other since 1979.[3] Prior to his return to India, he was working with the World Bank. During the period of attitudinal change in India he was additional and special secretary to the Prime Minister's office. Since then he has held positions in the Ministry of Finance, the Planning Commission, and the Economic Advisory Council to the Prime Minister.

Ahluwalia was the first director of the Independent Evaluation Office at the IMF in Washington, DC before he joined the Congress Party led Government in May 2004 as Deputy Chairman of the Planning Commission, of which the Prime Minister is Chairman. He was also the Finance Secretary of the Ministry of Finance when Man Mohan Singh was the Finance Minister and the principal architect of the reform process. Ahluwalia played a major role in the restructuring of industrial trade and fiscal policies that formed part of the first generation of liberalization reforms in India. He has been instrumental in driving the second generation of reforms that focus more on rapid growth with increased efficiency in the public sector's delivery of key social goods, such as education, health, water supply, and sanitation, with more effective provision of infrastructure. Throughout this period he has been a staunch advocate of private sector participation in various sectors including water reform. Private sector participation in infrastructure (including water supply and sanitation) is desirable not only to ensure a larger flow of resources but also to introduce greater efficiency in the supply of these services. The explosion of global capital markets and the associated expansion of private capital flows to emerging market economies provide new opportunities to finance infrastructure projects in such countries, if projects can be made commercially viable (Ahluwalia 1997).

Another important advisor in the key elite policy group has been Dr Rakesh Mohan. Holding an undergraduate degree from Yale and a doctorate in economics from Princeton University, he was inducted into the Government in the 1980s and has held key positions ever since. He conducted extensive research on urban economics, infrastructure development, and industrial and trade policy. As a committee member of the development policy program in the Planning Commission, economic advisor in the Ministry of Industry (1988–96), chairman of the National Council of Applied Economic Research, and economic advisor to the Finance Minister under the BJP Government, he was a key player in the process of liberalization. He headed several committees and was the chair of the Expert Group on Commercialization of Infrastructure Projects (1994–96) which published the *Indian Infrastructure Report: Policy Imperatives for Growth and Welfare* (EGCIP: 1996). This report made important recommendations for private participation in the water sector. Thus, the attitudinal change towards liberalization that began with Rajiv Gandhi's administration continued in other Governments precisely because of the continuity in the bureaucratic realm as well as the political support of the leadership during the period following Rajiv Gandhi's assassination.

The liberalization process in India differed from the standard approach of the "Washington consensus," a term coined by John Williamson to describe a set of ten economic principles that were designed as a standard reform package for crisis-ridden countries by the IMF, the World Bank, and the US Treasury Department – all Washington based institutions. The term is also used to describe a range of policies broadly associated with the rapid implementation of a policy that would expand the role of market forces and constrain the role of the state. In a plural and diverse democracy like India, the central Government opted for a gradual and controlled liberalization, reducing the speed of the reforms, which were sequenced and structured as the first generation and second generation of reforms in 1991.

The Rao period: the balance of payment crisis and the reform process in 1991

The reform process began in response to a developing economic crisis. In 1990, India's low Forex reserves forced the then finance minister, Madhu Dandavate, to announce to the world that India's coffers were empty. Consequently, access to foreign borrowing became difficult and India's credit rating was downgraded.

Under a loan negotiation with the IMF in July 1990, India accepted the structural adjustment program. Policy entrepreneurs started consultations with the IMF and negotiated a loan within a month of the Government's assuming office. In January 1991, the Janata Dal Government which succeeded the National Front Government, was forced to draw an additional IMF loan of $1.8 billion. In June 1991 India faced a severe balance of payment crisis as its Forex reserves plummeted to an all-time low of $1.2 billion, barely sufficient to pay for two weeks of vital imports (Ahluwalia 1994: 7). By 1991, the central Government's deficits had reached a record high of $17 billion, external debt had risen from $21 billion to $87 billion, and the debt-service ratio had increased to an unmanageable 32 percent of the gross domestic product (GDP) (Agarwal, *et al.* 1995: 159–66).

When Narsimha Rao became prime minister in July 1991, India was in desperate need of financial help and a substantial loan was a matter of necessity rather than choice. In a strategic move, Rao appointed Man Mohan Singh, an outsider to the ruling Congress Party and a long-term critic of India's export pessimism, as Finance Minister. His appointment was part of Rao's strategy to demonstrate continuity with previous policy, as well as to inspire IMF confidence regarding India's willingness and commitment to undertake structural adjustment programs (Dash 1999: 900). Singh's reputation as a neutral and effective financial manager during his previous appointments as Governor of the RBI and as secretary to the Ministry of Finance had earned him respect from all political parties. Thus, in Rao's calculation, Singh's initiatives for economic reform would not be subject to immediate partisan pressure, giving the minority Government a critical breathing space to mobilize support for the reform. Singh was also well acquainted with Michael Camdessus, Director General of the World Bank, with which the Rao Government was negotiating. Singh inducted Montek Singh Ahluwalia as his finance secretary. Both Singh and Ahluwalia planted "their men," all of whom had extensive World Bank or IMF experience, in most ministries (Dash 1999: 900), and this move helped establish a technocratic alignment between Indian bureaucracy and international financial agencies.

The Congress Party was sharply divided about the liberalization process. Some senior members were worried about the political repercussions of the IMF's market oriented policies and the Party's apparent abandonment of the pro-poor, populist orientation that had won them electoral victories in the past. The IMF's conditionality of external liberalization and privatization of the public sector units were not acceptable to them; however, they remained silent for fear of losing power. While Janata Dal and the BJP adopted a wait-and-see approach, which meant that they were not willing to make any official statement on their party's position on the situation at that point of time, the Communist Party of India (CPI

(M)) opposed the move. Ultimately though, the gravity of India's economic situation could not be ignored. Aware of the magnitude of the crisis – and in the absence of any real alternative for financial help – all parties had to accept the process of liberalization. India's economic crisis and the decision to seek a loan from the IMF thus set the country on a course toward economic stabilization and international confidence in the economy (Singh 1997: 16–42).

Business groups, on balance, supported the Rao Singh policy for economic liberalization. The prospect of new technology and capital appealed to the business community. In April 1991, the Confederation of Indian Industry (CII) prepared a theme paper on the contours of a free and competitive economy. In an effort to create a consensus on economic liberalization issues amongst various groups, the CII held numerous meetings with journalists, academics, trade union leaders, party leaders, and government officials including a first ever meeting with 227 parliamentarians from 17 states (CII 1991: 12).

The middle class became increasingly supportive of the Government's liberalization based programs which brought tangible economic benefits in the form of tax breaks and a higher personal income tax ceiling. As more foreign investment and opportunities for jobs with higher salaries led to more income and more purchasing power, the status-conscious middle class were able to buy quality goods available as a result of import liberalization (Dash 1999). The linkages that were forged among business groups, Government, and the middle class after the 1980s also help to explain the growing support of the middle class for economic liberalization policies.

Since the mid-1980s, India had been experiencing a quiet revolution, gradually moving from inward looking to outward looking economic policies. The crisis of 1991 facilitated this revolution (Pederson 2000). So in 1991, the Rao–Singh plan was sustained through (1) a small group of people involved in economic policy making, many of whom had previously been employees of the IMF or World Bank (Patnaik 2000) and (2) the rise of a new group of industrialists, committed to modern technology, professional management and collaboration with foreign companies that was mainly represented by the CII (Pederson 2000) and (3) a supportive middle class (Dash 1999).

Several factors came into play in the 1991 public acceptance of the liberalization process:

1 Rao's role as political maximizer and entrepreneur and his determination to create a coherent policy.
2 Singh's role as a perfect economic strategist.
3 Changes in the Indian political, social, and corporate landscape.

4 Links between the middle class, business, bureaucrats, and politicians.
5 Changes in cultural attitude.
6 The macroeconomic crisis of 1991, which offered an opportunity to undertake a politically risky program (Dash 1999).

The balance of payment crisis in 1991 thus created a window of opportunity to pass reforms for more extensive liberalization. The urgency mechanism provided the Rao Government the chance to work with different political parties to pass a sweeping reform package. Since the policy entrepreneurs were already in place with papers and draft policy documents, the extent of the reforms was not merely a response to the economic crisis. The state structures were already in place to facilitate these reforms. The political team of Finance Minister Man Mohan Singh, Commerce Minister P. Chidambaram, and Prime Minister Narsimha Rao helped the policy bureaucrats working on the liberalization program. The role these leaders played provided the space for technocrats to develop details of policy while protecting them from public eye.

The IMF, for its part, wanted to undertake fiscal consolidation by agreeing to a set of terms that would go along with its stabilization and structural adjustment program, which meant reduction of fiscal deficits, balancing the budget, cutting subsidies, and increasing food and fertilizer prices. In fact, the budget of 1991–92 had to cut expenditures drastically, and the axe fell on social sectors and capital expenditures (Nayar 2001). India also applied for a loan from the World Bank. The loans that India took all came with conditions. The larger policy of economic reform occurred simultaneously with economic stabilization and structural adjustments – a process of attacking the more fundamental features of the economy such as the relationship between the market and the state and the national and world economy, as well as shrinking the state in favor of the market, both internally and externally (Kahler 1990).

In the case of India, it may seem logical to some that both the ideas and interests model, which included a significant minority of the elite and politicians who shared the assumptions of the new orthodoxy, and the power model of the international institutions and their coercion initially became the sources of reform in India. Through a process of social learning, key leaders came to the understanding that earlier policies had not met their goals, reflecting the changes adopted by Indira Gandhi in 1974 and Rajiv Gandhi in the 1980s. The bureaucratic elite were also convinced of the need for change. From a domestic viewpoint, the key variables can be understood to be the state, with its institutions; the elite and policy preferences (Haggard and Kaufman 1992: 3–37); and the pressures of an economic crisis. Thus, in the words of Nayar, "If the IFIs [international financial institutions] applied pressure at all, they were pressing against an already open door" (2001: 146).

The process of liberalization in the Indian state was not accomplished by the IFIs putting into place a generalized menu of international economic regimes that advanced the interests of their multinational corporations' search for new markets. In the Indian case, political policies were filtered through the political leadership's own assessment and evaluation of what reforms would be feasible for building a consensus and what would be acceptable to the given population at a given time. The democratic nature of the Indian state and India's long tradition of political democracy required a more cautious approach. The main idea behind the reform was not to end the role of the state as a protector of the vulnerable and promoter of development, but rather to preserve this role as much as possible. The reforms were centrally managed and this management, including the sequencing of the reforms, was done very carefully. While the reforms were publicized in international forums, their importance was underplayed in domestic politics. Narasimha Rao realized that a low key approach would help to reduce opposition to the reforms (Manor 1995).

The state needed to balance the concerns of legitimacy and accumulation of capital where the nature of sacrifices imposed on key social groups limited the type and scope of acceptable changes. Thus, the reforms were sequenced over a period from one budget to the other. Those pushing for liberalization clarified to the World Bank and the IMF that while India was willing to accept their conditionality, it would phase the changes in accordance with the political climate of a minority Government and the public at large. The politically acceptable changes that formed part of the soft reforms were implemented immediately. However, the hard reforms of subsidies; exit policy; privatization of public sector enterprises; ending government subsidies; and power, insurance, and water sector reforms were phased in when a window of opportunity opened, under a more gradual and incremental approach. While hard reforms were kept in abeyance for the right opportunity and political consensus, preparations were underway not only within the administration but also in the IFIs to address issues facing the critical and highly controversial water sector. Before reforms were enacted, domestically high powered committees were set up, like the Tax Reform Committee, the Committee on Financial System, the Insurance Reform Committee, and others. One of these was the already mentioned Expert Group on Commercialization of Infrastructure Projects (1994–96) whose scope included water supply and sanitation. This committee (discussed in detail in Chapter 4) made strong recommendations for private sector participation in water supply and sanitation.

This period saw a significant shift in the state's overall logic of economic development in its acknowledgement of internal administrative reform coupled with a new faith in foreign investment and global competition (Bhaduri and Nayyar 1996). The Indian state was no longer wedded to the Nehruvian state – intervention and the development vision had lost its legitimacy among the elite.

Narsimha Rao's coalition completed its term in 1996 and was followed by a United Front Coalition that formed a Government on 1 July of that year. This coalition was supported by the Congress Party and consisted of several regional parties and the Left Front, all of whom came together on the issue of a secular Government in opposition to the BJP, which had a more fundamentalist ideology. The reform process continued under the United Front Coalition Government. The reforms concentrated on fiscal management; foreign direct investment and priva- tization; disinvestment of public sector enterprises; and lowering of tariffs. These reforms were largely restricted to the center and therefore touched only those items that came under the Union List and the Concurrent List (see Chapter 4). An important initiative of the coalition Government was the establishment of the Telecom Regulatory Authority as an independent agency to regulate all telecom- munications operators, both government and private.

The BJP and economic liberalization

The United Front Coalition could not complete its full five-year term, and in the next elections the BJP coalition came to power in 1998. While in opposition, the BJP, in an uneasy alliance with the Left, led several campaigns against a wide range of Western corporations. But once in power, the Government continued on the path to reinvent the role of the state in relation to economic development with a vision of global integration. Despite its ideological focus on "Swadeshi" and "Economic Nationalism," the party's conversion to the ideas of globalization and liberalization was manifest in Finance Minister Yashwant Sinha's statement after taking charge (*Times of India* 21 March 1998). Saying that "the 1991 reforms were a step in the right direction" and that "multinationals had nothing to fear" from BJP's policy of economic reforms, Sinha vowed to deepen, broaden, and accelerate reforms (*Times of India* 26 March 1998). In a major policy pronounce- ment R. K. Hegde, the Minister of Commerce, inaugurated the second generation of reforms in April 1998 by boldly moving to phase out quantitative restrictions on imports and hastening the process of integrating the national economy with the global economy he labeled as "Swadeshi Liberalism" (*Times of India* 15 April 1998). This was followed by Power Minister P. R. Kumarangalam's bill to set India's state owned electricity supply industry in order. The bill aimed to convert the state electricity boards (SEBs) into an independent state regulatory commission which would remove them from the control of individual state Governments and rescue them from their bankrupt positions by establishing commercially viable power tariffs. The bill was an indicator of the reformist intent of the Government of India and its cautious entry into the soft sector reforms. Although it was intended to be a mandatory measure, the bill was made optional and left to the discretion of individual state Governments due to the opposition of some coalition parties. These parties did not want any interference in the political practice of supplying highly subsidized or free electricity to farmers. Additional actions included passing the Urban Ceiling and Regulation Act (1999) and opening up the insurance sector to private competition by allowing foreign companies 40 percent equity. These reforms continued till the BJP completed its full term in office in 2004. It was during this period that the National Water Resources Council in the Ministry of Water Resources met to review the National Water Policy of 1987 and draft a new water policy for India in 1998, in view of the larger changes in the Indian economy.

Changes in water policy reform now seemed to be imminent. Narsalay (2003: 3) sums up the state of this reform process:

> Internationally, as 'economies' started eating into the space of societies and as different elements of the structural adjustment programs started gaining political acceptance as the only macroeconomic answer to achieve developmental goals, a strong political pitch started being made even in India with respect to issues in realm of ownership and rights, over natural resources including water.

Instead of galvanizing a set of political strategies that could encourage and empower people to make the state more efficient and responsive, or to make the people more central in the planning process, the proponents of the new thought process argued that it made more sense to hand over the reins of the new power structure for deciding water rights to private agents. The following chapter details the shift in India's water policy as a result of both internal and international processes of liberalization in the context of India's reform policy.

Notes

1 The model aimed at transforming India from an agricultural economy into an industrial one. It was a command economy based on centralized development planning; extensive ownership of commercial assets; a complex industrial licensing system; substantial protection against imports; restriction on exports; and virtual prohibition

of foreign investments. See http://www.tribune India.com/50yrs/kapur.htm (accessed 11 September 2007).

2 Rajiv Gandhi's Government gained 77 percent of the seats in Lok Sabha in the elections that followed his mother's assassination by her bodyguards.

3 The Deputy Chairman of Planning Commission in the Man Mohan Singh led coalition Government (2004–2009).

4   Water in the liberalization process

The processes at work during the period of water policy reform (1991–2005) were multidimensional and multifaceted. The roots of their complexity reach far back into India's history and spread throughout many of the institutions that define India as a nation. In order to understand these processes fully, we must go back to ancient times as well as travel through the era of British colonialism; we must understand how India's constitutional system shapes the context of water policy development and implementation. We must also understand the institutions and actors that shaped urban water policy, the National Water Policy of 1987, and the external influences at work in shaping the National Water Policy of 2002, as well as their impacts on state policies in general.

Water in ancient and British India

A number of studies carried out by historians of ancient India[1] have shown that water management from ancient times had been in the hands of local society. Water management of innumerable water works in villages and the countryside in the form of dams, tanks, wells, reservoirs, lakes, step wells, etc. were managed by local people (Allchin 1998). The local control over water and land paved the way for development of the social, economic and political autonomy of villages and communities and regions that often negotiated with, and resisted the authority of, the centralizing control of state or empire. Water was managed through a system of patronage and community control through village councils in most parts of India. Prior to the arrival of the British in South India water was collectively managed by communities through a system called "kudimaramath" (self-repair).

The advent of the corporate rule of the East India Company[2] in the eighteenth century brought about a major shift in policy over collectively managing resources through public funds. Before then peasants had paid 300 out of 1,000 units of grains to public funds and 250 of these units stayed in the village for maintenance of commons and public works. By 1830, peasants started to pay 650 units out of which 590 units went straight to the East India Company. As a result of this increased siphoning of payments and loss of revenue in public funds, the peasants and commons were ruined. Some 300,000 water tanks built over centuries in pre-British India were destroyed affecting agricultural productivity and earnings (Shiva 2002).

Even though public works in water and irrigation were found to be profitable, the Company was reluctant to expand these works further. Irrigation was accorded a very low priority in Company policy (Bhattacharya 1975: 40). However, due to pressure by nationalists, the Company in 1857 took steps to construct irrigation works. Just like railway construction in British India, the irrigation projects were to be funded with private capital. In 1857, the Court Directors of the Company invited a proposal for private construction of irrigation works in Madras (*Imperial Gazette of India* 1908). The first private construction was the Tunghbhadra project undertaken by the Madras Irrigation and Canal Company with 5 percent government guarantee on the capital investment. Within nine years the company faced acute financial crisis and the project was taken over by the Government in 1882 and absorbed into the public sector (Bhattacharya 1975: 41). During this period the Company did invest some private capital in water but the failure to sustain that process led to the absorption of the company into the public sector.

The colonial state that took over India in 1858 systematically broke the backbone of local autonomy. First, it drastically reduced and then discontinued the state allocation of capital for maintenance of local water works by the introduction of definite property rights in land; it then imposed a highly exploitative land revenue system and had total control over all the natural resources of the subcontinent. The intention was to maximize revenue by the commercialization of land, forests and, most importantly, water (Satya 2001). With the withdrawal of state patronage, local water works that had existed for time immemorial fell into disrepair, disuse, and completely disappeared from large parts of South Asia.

The colonial rule sponsored irrigation works in the form of canals and barrages, dams in Punjab, the United Provinces, and the North-West Provinces. The Bihar, Bengal and Madras presidencies were designed to promote commercial crops such as cotton, opium, jute, indigo, sugarcane, tea, coffee, tobacco, wheat, etc. instead of food grains. Water resources were under state control and were meant to keep India's economy serving British interests. The colonial desires were to (1) control water and other natural assets to legitimize the empire; (2) exploit the natural treasures for profiteering; and (3) send wealth from India to the British Crown. The public irrigation works yielded substantial profit and they were looked upon as successful projects constructed during days of the Company (Edwardes 1967: 93).

In this context, it would not have been in the interest of the British to allow private enterprise to develop in India. So all infrastructure building specifically in water was British controlled and meant to serve the imperial needs of the British economy. In fact, Indian revenues were servicing British debts to the United States and Germany in the nineteenth century, irrespective of the fact that millions were dying in absolute poverty, squalor, disease, and famine. Private investments in general were discouraged by the colonial state, and it was especially discouraged in the augmentation, management and distribution of water in either the rural areas or towns, because water was one of the principal sources of revenues for the colonial state. Privatizing water would mean the loss of this precious source, which the British Government in India would never have allowed to happen. Thus, the colonial state encouraged the private sector in cotton, jute, and cash crops, but retained control of infrastructure like railroads, telegraphs and water resources in the public sector which maximized its revenue.

During the second half of the nineteenth century there were canals constructed under public and private enterprises for irrigation purposes. The large canals were mainly built by the Government but private initiative was not negligible. Under private construction were 1,000,000 acres in Punjab, 750,000 acres in the United Provinces, 5,000,000 acres in Bengal and 336,000 acres in Madras. However, taking estimates of the total area under private irrigation in India into account, the annual figure was about 8,000,000 acres, which was about 16–18 percent of the total (*Imperial Gazette of India* 1908: 325). The colonial state and the imperial mentality in the development of irrigation in the Indus Basin saw an alliance between engineering sciences and the state based on a technocratic vision of harnessing and controlling nature to maximize revenue (Gilmartin 1994). Thus, the private sector was never encouraged to invest in water by the British while they were in India. Private enterprises and capital was used in building canals but water management, control and distribution in both rural and urban areas was totally under state control.

Water in post-colonial India

The independent, post-colonial Indian state developed along a Nehruvian vision of socialism. Like pre-colonial and colonial India, the state was a highly centralized state in terms of economic management and political management. Five-year plans were floated on the lines of a Soviet model of development that gave primacy to addressing the needs of food and water security in agriculture and water distribution. The post-colonial independent India opted for the construction of large dams and hydroelectric projects such as Damodar Valley, the Bhakra Nangal Dam, Hirakund, Silent Valley, Narmada Valley and the Tehri Dam to solve the problems of water scarcity, insufficient drinking water supplies and energy crises. These projects were mainly state owned but were completed with the help of private capital. Urban water supply, management and distribution were in the hands of the public sector. The traditional methods favored under the Gandhian model and the ideals of localized control of management of water and land were rejected in favor of the so-called "Temples of modern India" (Nehruvian India). Control over water resources remained with the public sector and the constitution specifically allocated urban water supply to individual states.

Water in the Indian constitution

As a union of individual states, India's constitutional provisions with respect to the allocation of responsibilities between the states and the center fall into three categories:

1   Union List (List I), which empowers the center to legislate on the subjects mentioned in that list.
2   State List (List II), where individual states are autonomous in legislating on the subjects mentioned in that list.
3   Concurrent List (List III), where both the center and the states are empowered to legislate on the subjects in that list, but in the event of a conflict, the decision of the center prevails.

In the Constitution of India, "Water" is included as Entry 17 in List II and therefore constitutes a state legislated subject. This entry is conditional on the provisions of Entry 56 of List I, i.e. the Union List.

Entry 56 of List I of the Seventh Schedule states "Regulation and Development of Inter-state rivers and river valleys to the extent to which such regulation and development under the control of the Union is declared by Parliament by law to be expedient in the public interest". Entry 17 under List II of Seventh Schedule provides that "Water, that is to say water supplies, irrigation and canals, drainage and embankments, water storage and water power are subject to the provisions of Entry 56 of List I" (National Water Policy 1987).

Water is essentially a state legislated subject with the center having minimal intervention in its policies unless it becomes a matter of public interest. It was only in 1985 that a Ministry of Water Resources was established due to the growing pressures of diverse water issues. The National Water Resource Council was set up under Prime Minister Rajiv Gandhi under whom India adopted its first National Water Policy in 1987.

The Ministry of Water Resources is responsible for the coordination, development, conservation, and management of water as a natural resource. The ministry looks into the general policy on water resource development and management, manages technical and external assistance to the states for irrigation, multipurpose projects, groundwater exploration and exploitation, command area development, drainage, flood control, waterlogging, sea erosion problems, dam safety, and hydraulic structures for navigation and hydropower. The Ministry of Environment and Forests oversees water pollution and control issues while the Ministry of Power manages hydropower. The Ministry of Urban Development handles water supplies and sewage disposal in urban areas. Water supply in rural areas is taken care of by the Ministry of Rural Development.

Water being under the control of the states, the state Governments are primarily responsible for the management of this resource. The administrative control and responsibility for managing water rests with the various state departments and corporations. Urban water supply is generally the responsibility of the municipal corporations and water boards constituted under an Act of the state legislative assembly. These boards are autonomous in their functioning. Currently water supplies, from bulk production to distribution, are taken care of by these corporations and boards. The cost of water is highly subsidized by the state agencies as water is perceived as a public good. The rural water supplies are taken care of by the Panchayats. Urban Local Bodies (ULBs) get funds generally in the form of loans or grants from the central and state Governments. These local institutions, which are primarily responsible for urban water and rural supplies, were recognized as a third tier of the Government by a constitutional amendment Act in 1993. Under the 73th and 74th Amendments to the Indian Constitution these municipal corporations and Panchayats are mandated and empowered to chart out the financial and political character of development on issues like water supplies and management. Focusing as it did on "decentralization" in the Indian political system, the amendment was of such a magnitude that it had an impact on the functioning of the political machinery associated with water at the level of states and sub-state levels.

It is within this administrative set-up that policies for water resource management in irrigation as well as rural and urban water supplies are produced, enacted, and implemented. India adopted its first National Water Policy in 1987 under the leadership of Rajiv Gandhi in an era when ideas for the transition of the Indian state were actually being germinated. Perhaps that is why the policy makers, according to critics, "either wanted to keep covering their policy positions with a robe of ambiguousness or were naïve to the emerging political realities of water markets in future" (Narsalay 2003: 4). Some of the key features of the policy were:

- Water is a precious national resource and national perspectives should govern its development.
- In the allocation of water, ordinarily first priority should be for drinking water, with irrigation, hydropower, industrial uses and other uses following in that order.
- There are complex problems of equity and social justice concerning water distribution.
- Water rates should be such as to foster motivation for economy of use and should cover maintenance and operational charges and a part of the fixed costs. (National Water Policy 1987).

Critics commented that the policy was ambiguous as it was unclear whose "national perspectives" it referred to and if they essentially reflected the interest of the marginalized (Narsalay 2003). The 1987 policy makes no mention of private sector participation in the water sector.

The processes of water policy reform and the shift in the water paradigm need to be understood in this context. The first generation of reforms (1991–96), as discussed in Chapter 3, focused on issues relating to the economy. The second generation process of reforms (1998 onwards) focused more on the social and welfare sectors and were gradual and incremental. The idea behind this approach came from the working of the Indian state, where the political and ideological realities resist fast changes. The socialist ideology in India had deep roots stretching back nearly four decades and so policy entrepreneurs recommended the "go-slow process" in reforms. For nearly a decade after the onset of economic liberalization in India, a key component – privatization – remained dormant. The usual explanation has been that weak Governments could not overcome the many vested interests, from rent-seeking bureaucrats and ministers to public sector trade unions. The strategy, however, was to keep working by preparing detailed papers and committee reports pushing for change whenever a window of opportunity opened. Thus, when the Indian president in his opening address to Parliament in the 2002 budget session stated, "It is evident that disinvestment in public sector enterprises is no longer a matter of choice but an imperative. The prolonged fiscal hemorrhage from the majority of these enterprises cannot be sustained any longer," Indian privatization started to emerge out of the shadows.

Water in the reform process: 1991–2000

The first generation of reforms (1991–96) created a policy space for private agents to experiment with ideas that were essentially aimed at transferring decision making rights on ownership, pricing, use, and preservation of natural resources like water and services. The goal was to move policy making from an entirely public domain to the semi-private domain, if not privatizing completely. In the earlier years of the reform period, water sector reform did not figure in the policy discourse as water was seen as a politically sensitive subject and a public good. But preparations were already under way and a high powered, expert committee under the chairmanship of Dr Rakesh Mohan was set up. The committee released its report entitled *Indian Infrastructure Report: Policy Imperatives for Growth and Welfare* in 1996 with a detailed focus on water supply and sanitation.

Report of the Expert Group on Commercialization of Infrastructure Projects

The report cited major problems in the water supply sector such as low water tariffs, high costs of production, high system losses from supply, poor demand management, and low cost recovery. According to the report, informal private players had taken over water supplies, but this had attacked the monopoly position of the local authorities only marginally. Being in the public sector, the authorities did not take competition as a threat at all: efficiency and cost recovery did not improve. High administrative costs of establishment, because of wages and salaries, continued to account for a major chunk of the costs (EGCIP 1996). Citing the example of developed countries, the report said "With a standard tariff round the year, the concept of water conservation in summer is absent in India, leading to a high demand and low return situation" (EGCIP 1996: 30). The report recommended unbundling services and the involvement of public and private players.

> Public private partnerships (PPPs) may be conceived as a first step towards privatization pending legal reforms (permitting entry of to the private sector) and institution of a regulatory framework. This may also be the feasible mode of privatization.
>
> (EGCIP 1996: 17)

Advocating the entry of private players into the water supply system, the report recommended public–private partnerships (PPPs) in the various individual components of water supplies or for the full system. The water supply system has the following components: (1) raw water source (2) transmission of raw water and

(3) distribution. Presently the whole operation from source to distribution is taken care of by the water supply boards. Citing the United States as an example, the committee recommended that the ideal solution would be that the boards produce bulk treated water and privatize the retail operations. "We suggest the contracting out of part or the whole of the service" (EGCIP 1996: 31).

The report stated that improving cost effectiveness through competition, technological upgrading, differential levels of treatment, and price and cost recovery by increasing water tariffs, connection charges, registration charges, and betterment charges could improve efficiency and rationalize consumption. Regularity in supply could mean lower project costs and a greater willingness to pay on the part of the consumers.

The report also suggested that the ULBs responsible for water supplies should be entrusted with the functions of planning coordination and developing policy for the supply of services. The ULBs would be responsible for contracting out operations, preparation of contract documents, and monitoring of private operators.

> Unbundling of services has been a major mechanism through which the misconception about the economies of scale argument has been overcome. Governments have not been very successful as effective suppliers of many services. Commercialization of infrastructure projects means efficient provision of services to the consumers' satisfaction on cost recovery basis. Since the public sector in most cases is an inefficient provider due to its inherent characteristics, promotion of privatization itself becomes an instrument of commercialization.
>
> (EGCIP 1996: 43)

The report thus laid emphasis on efficiency, cost recovery, decentralization, and governance in water supplies and sanitation policies.

One needs to take into consideration that this report was released in a period that witnessed tremendous political uncertainty at the center, with minority Governments and unstable coalitions.[3] While there was continuity in the reform process, these conditions actually translated into ambiguous political signals with respect to privatization, liberalization, and globalization of infrastructure services, including water supply. This ambiguity, coupled with a lack of political consistency, not only affected the quality of investment in the water sector, but also caused the growing concerns over natural resources to be ignored. Water, a sensitive subject, was not in the forefront of the national agenda. But planning for changes in the National Water Policy of 1987 was imminent with the release of the draft National Water Policy in 1998 and certain international developments during this period.

International developments during the period

The pace at which the dominance of neoliberalism as a discourse or ideology at the global level circulated in the 1990s had important consequences for water at the national level in India. This discourse of globalization asserted that water was to be distributed by mechanisms of the market. "Over the last 20 years no global water policy meeting has neglected to pass a resolution which, among others, defined water as an 'economic good'" (Wolf 2003: 174). While policy shifts towards privatization and commercialization of water services in developed economies are often based on national decisions and regulations, developing countries are increasingly subject to international commitments compelling the implementation of privatization measures. The IMF, the World Bank, and the regional development banks have played a key role in the restructuring of public owned services, including privatization of the water sector in low income countries, a condition of loans and debt relief. The reorientation of the role of public and private entities in the water sector has also been reinforced as a result of the development of international trade law whereby the liberalization of trade in goods has been joined by negotiations toward the progressive liberalization of trade in services under the World Trade Organization (WTO) and its General Agreement on Trade in Services (GATS). The aim of GATS is to progressively liberalize trade in services, removing restrictions and internal government regulations considered barriers to trade in the area of service delivery, including education, health care, tourism, transport, waste collection, and water.

These developments on the international level and the formation of the WTO had strategic implications for water policy at the national level in India. Privatization of the water supply became high on the political agenda after developing countries accepted the GATS, which is one of the cornerstones of the WTO regime. When water is seen as a global commodity, the control of natural resources for profit is encouraged. This posits a major paradigm shift in the use and access to these life resources that transforms water into an object of trade with a price determined by the logic of market competition and profit.

India became a member of the WTO in 1995 and a signatory to the GATS. The Indian water policy needed changes, which led to the formation of a committee under the National Water Resource Council (NWRC) in 1998 that produced a draft of the National Water Resources Policy. However, since there was little awareness among political and business groups on issues pertaining to GATS/

WTO, the issues relating to water never really surfaced during the ongoing negotiations in the Uruguay Round.

Meanwhile, the World Bank, which had been the largest donor to India
since the 1950s on different water projects, produced two reports on India: the
Irrigation Sector Review (1991) and the *Water Resources Management Policy*
(1993). The World Bank highlighted that severe organizational and instructional
problems persisted despite the adoption of the National Water Policy (1987).
Realizing that project-by-project assistance did not work and project loans had
become a disbursement exercise, the World Bank decided to switch from project
to sector loans in the area of water. The report emphasized that water is a "scarce
commodity" and argued that "the scarcity value of water" must be reflected in
water charges. Another aspect it emphasized was that "affordability" would not
be an issue in "efficient" water management in India.

An important strategy advocated in the World Bank's *Water Resources
Management Policy* was to reduce the role of the Government from being the
sole provider and financier to being a facilitator, enabler, and regulator. The
report asserted that sustainable growth could be achieved only by improved
water management, which would result from overcoming constraints posed
by the quantity and quality of water available for development. The failure of
these past practices to adequately deal with the challenges highlighted the need
for a new thinking in the World Bank's 1993 report, termed "sustainable water
resources management."

Arguing for institutional strengthening and reorganization, the reports
advocated a shift from the supply driven to a demand driven approach with
an appropriate framework to separate policy and regulatory functions from
operations.

The World Bank advocated the need for change on the basis of "poverty alleviation," claiming that "the poor are much better off when water is managed as
an economic good" (Briscoe 1996: 3). According to the World Bank, increasing
prices to enable cost recovery in the delivery of services may actually help the poor.
The argument is that the poor often pay higher prices to private vendors, as they
are not connected to public services networks. Moreover, easier access to water
can free up time, which can be used to earn income and other productive uses.

Unlike the earlier project-by-project investments, the new approach was
to cover water resource management, address current and future intersectoral
needs, and provide support for institutional development and reform in line with
the World Bank's policy. The World Bank also brought out a series of publications in the 1990s that re-emphasized the reduction of public sector intervention,
ensuring appropriate costs for infrastructure through elimination and reduction

of subsidies. It stressed the development of capital markets for resource mobilization, facilitating private and joint sector projects using PPPs to enhance efficiency. These publications included *Reducing Poverty in India: Options for More Effective Public Services* (1998); *India: Urban Infrastructure Services Review* (1996, 1997); and *Urban Water Supply and Sanitation* (1998). The goal of the World Bank was thus to reduce monopolies while subjecting everything else to market mechanisms in infrastructure development, particularly water. Table 4.1 provides an overview of some of the reports that the bank produced on India's water sector.

In *India: World Bank Assistance for Water Resources Management – A Country Assistance Evaluation* (2002), the World Bank stated "A culture of 'government must do it' prevails, and that the sector's bureaucracy has grown unwieldy, not adaptive to changing needs with narrow interests and lacking incentives to improve performance." It recommended that both state and center reassess their monopoly and ascribe new roles as necessary and that "where appropriate", and "where opportunities abound," they should divest as much investment and implementation as possible to the energetic nongovernment sector (Pitman 2002: 17).

Adopting the concept of "unbundling," a term first used by neoclassical economists, the World Bank endorsed it by stating that "By isolating the natural monopoly segments of an industry, unbundling promotes new entry and competition in segments that are potentially competitive" (World Bank 1994: 53). All these reports emphasized the need for reforms in the water sector to correct deficiencies, provide services, and improve management and performance of the sector (Singh 2004: 60). Bureaucratic visits from the Ministry of Urban Development and the Ministry of Water Resources also followed at the headquarters of the World Bank in Washington DC.

The World Bank and the Public–Private Infrastructure Advisory Facility (PPIAF), a multidonor agency for private sector participation in infrastructure, also organized workshops and training programs for journalists in environment, PPPs, and public sector reform to disseminate information about the relevance of the policies of these international financial agencies. The PPIAF launched a major initiative through a consultative workshop of policy makers and stakeholders on 31 October 2000, known as the Water Policy Reform Initiative – India, with a grant of $520,000 by the PPIAF, and $430,000 from co-donors like the World Bank and Sain International. The objective of the initiative was to alleviate poverty by building consensus on water sector reform in India, and strengthening the capacity of decision makers and stakeholders to prepare and implement reforms leading to increased private sector participation in the water sector. It

Table 4.1 World Bank reports on India's water sector: an overview

	Year	Report title
1	1990	*Irrigation Sector Review – A Review of the Bank's Past Strategy in the Irrigation Sector Review*
2	1992	*Comprehensive Water Resource Management: Concept Paper*
3	1993	*India: Water Resource Management Policy*
4	1995	*India – Private Infrastructure (IL&FS) Project: Environmental and Social Report*
5	1996	*India – Private Infrastructure Finance Report*
6	1998	*India – Water Sector Management Sector Review: Initiating and Sustaining Water Sector Reforms*
7	1998	*India – Water Sector Management Sector Review: Groundwater Regulation and Management*
8	1998	*India – Water Sector Management Sector Review: Irrigation Sector*
9	1998	*India – Water Sector Management Sector Review: Urban Water Supply and Sanitation Report*
10	1998	*India – Water Sector Management Sector Review: Rural Water Supply and Sanitation*
11	1998	*India – Water Sector Management Sector Review: Report on Intersectoral Water Allocation, Planning, and Management*
12	1998	*Fifth Meeting of Urban Think Tanks on Financing Options for Water and Sanitation: Working Paper*
13	1998	*Sixth Meeting of Urban Think Tanks on Institutional Arrangements for Provision of Water and Sanitation to the Poor: Working Paper*
14	1999	*Water Challenge and Institutional Reform: Cross Country Perspective*
15	1999	*Eighth Meeting of Urban Think Tanks on Building Municipal Capacities to Deliver Services to the Poor: Working Paper*
16	1999	*Ninth Meeting of Urban Think Tanks on Private Sector Participation in Provision of Water and Sanitation Services to the Urban Poor: Working Paper*
17	1999	*Water for India's Poor – Who Pays the Price for Broken Promises?*
18	2001	*India – Country Assistance Evaluation – Operation Evaluation Study*
19	2001	*Twelfth Meeting of Urban Think Tanks on Tariffs and Subsidies: Working Paper*
20	2001	*Launching Sector Reforms – Government of India Pilot Demand-Responsive Approaches to Rural Water Supply and Sanitation*
21	2002	*India – World Bank Assistance for Water Resource Management: A Country Assistance Evaluation*
22	2004	*Institutional Reform Options in the Irrigation Sector*
23	2004	*India – Attaining Millennium Development Goals in India – Role of Public Policy and Service Delivery*
24	2004	*India – Urban Finance and Governance Review Vols. I & II*

(continued)

Table 4.1 (cont.)

	Year	Report title
25	2005	*India – National Urban Infrastructure Project*
26	2005	*India's Water Economy: Bracing for a Turbulent Future*
27	2006	*Water Supply and Sanitation – Bridging the Gap between Infrastructure and Services*

Source: Compiled from http://www.worldbank.org

would also promote knowledge sharing on reform and institutional development for improved sector performance and better meet the needs of the urban poor – through greater awareness of the rationale for, and best practices in, such reform (PPIAF 2003).

To achieve this objective, the Water Policy Reform Initiative relied on three key elements:

1 **Policy dialogue.** This dialogue consisted of policy seminars for state level decision makers and stakeholders focusing on the main barriers to reform. The goal was to promote consensus at the state level for policy reform, as indicated by the outputs and declarations produced by workshops, seminars, and presentations which included an Urban Water and Sanitation Sector Reform Workshop: Piloting Private Sector Participation in Mega Cities; the Twelfth Meeting of the Urban Think Tank on Tariffs, Subsidies, and the Poor in the Indian Water Sector; an International Conference on New Perspectives on Water for Urban and Rural India; and a conference on Private Sector Participation in Urban Water and Sanitation Services: Managing The Process and Regulating the Sector.

2 **Public awareness.** Information seminars for journalists and members of civil society were organized to influence public opinion, as measured by the number and quality of articles and other media reports following journalist workshops and the advocacy efforts undertaken by think tanks, nongovernmental organizations (NGOs), and the like. The Initiative organized "Running Water: A Dialogue for Journalists" to build an informed press to improve coverage of the water sector.

3 **Knowledge product production and dissemination of knowledge products.** To support the initiative, knowledge product was developed in the form of a series of tariff and subsidies papers and household surveys in selected cities. The papers were distributed to relevant policy makers, service providers, and other stakeholders in India and the rest of the region. These reports were posted on the Ministry of Urban Development websites for

like-minded states and local utilities to have easy access to the information (PPIAF 2003).

Goldman's (2005:194) view best sums up these initiatives.

> The idea of contracting out public goods and service provisions to the private sector, and in particular to globally competitive bidders, becomes more than an ideological fantasy but a "best practices" case that gets explored in the classroom, with experts flown in to demonstrate its utility and viability, and then gets realized in the field through development projects.

During this period, recommendations of the National Water Policy (1987), the World Bank's Irrigation Sector Review (December 1991) and the Joint Government of India/World Bank *Water Resource Management Sector Review* (June 1998) were targeted for implementation. The most important achievement of the World Bank during this period was the way in which it deepened its institutional understanding of the bureaucracy and of the polity and biases that plague the political economy of water at the national and subnational levels in India. Understanding the political, sociopsychological, and cultural dimension of water in India, the World Bank began to champion the cause of private sector participation couched in the discourses of *crisis, scarcity, poverty alleviation, and statistics.* These discourses and internal liberalization measures led to the official publication of important policy documents in water that reflect the shift in the national agenda for water policy reforms in the early years of the twenty-first century.

Policy documents in the agenda for water policy reform: 2000–2005

As the second generation of reforms began in 1998, the draft National Water Policy (1998) was adopted on 1 April 2002, amidst a combination of internal liberalization measures and external pressures. The water policy marks a departure from the 1987 policy in several ways. It lays emphasis on socioeconomic aspects in water policy planning and the needs of the individual states. The addition to the policy of Sections 11, 12, and 13 reflects the reformist intent of the Government in a neoliberal framework. Without defining what "private sector" means in the context of the policy, the document asserts:

> Private sector participation should be encouraged in planning, development and management of water resources projects for diverse uses, wherever

feasible. Private sector participation may help in introducing innovative ideas, generating financial resources and introducing corporate management and improve service efficiency and accountability to users. Depending upon specific situations, various combinations of private sector participation in building, owning, operating, leasing and transferring of water resource facilities may be considered.

(National Water Policy 2002)

The fact that this document explicitly encourages "corporate management" as one of the roles of the private sector naturally provides the space to push for corporate control over the resources. Community management does not figure in the policy. The word "community" is used only once in the Conclusion, mentioning that the "concerns of the community need to be taken into account for water resource development." So powerful were the vested interests, that the policy did not incorporate Prime Minister Vajpayee's stated view at the Fifth Meeting of the National Water Resource Council that National Water Policy should be people centered and recognize communities as the "rightful custodian of water" (SANDRP 2002). Sunita Narain, Director of the Centre for Science and Environment, New Delhi, had vigorously campaigned for rainwater harvesting and community rights in water and declared that "The National Water Policy will remain inert and ineffectual because it is far removed from the two simple but important challenges of water management today – rainwater harvesting and community management in this initiative" (Narain in Devraj 2002).

On the participatory approach to water resource management, the policy clearly states that:

Management of water resources for diverse uses should be done by adopting a participatory approach; by involving not only the various governmental agencies but also the users *and other stakeholders* [my emphasis], in an effective and decisive manner, in various aspects of planning, design, development and management of the water resource schemes.

(National Water Policy 2002)

Who constitutes a legitimate stakeholder in any decision making process? What exact rights do these stakeholders have in terms of information prior to the decision? The role they have in the decision making processes is vaguely expressed and ambiguous in the policy.

On financial and physical sustainability, the policy declares:

Adequate emphasis needs to be given to the physical and financial sustainability of existing facilities. There is therefore a need to ensure that water charges for various uses should be fixed in such a way that they cover at least part of the capital costs subsequently. These rates should be linked directly to the quality of service provided. The subsidy on direct water rates to the disadvantaged and poorer sections of the society should be well targeted and transparent.

(National Water Policy 2002)

Cost recovery thus becomes an essential goal of the new water policy. The policy also states that to achieve the desired objectives, "state water policies backed with an operational plan shall be formulated in a time bound manner, say in two years." Thus, with the signing into law of the New National Water Policy by Prime Minister Vajpayee in the National Water Resources Council meeting (1 April 2002), citizen accountability was transformed into customer choice. The general reaction to the policy was expressed in the words of the Rashtriya Jal Biradari (the National Water Community), a coalition of NGOs working in water: "Water will be privatized with transnational corporations managing access to it on the basis of profit" (Rashtriya Jal Biradari 2002). Critics commented that the emphasis of the policy was on centralization, expert oriented guidance, and participation of private players (Singh 2004). The objective of the National Water Policy (2002) was not to achieve the universal availability of water across and within sectors or to make all the completed water infrastructure projects subject to performance review and the place the reports in the public domain with consultation (Vombhatkere 2005). The policy also does not reflect the priorities and concerns for social justice or the environment in the actual policy measures that the document puts forth. There is a national call to review the National Water Policy and to frame a new policy through a nationwide consultation that must include clearly defined policy about transparency, accountability, and participation in planning, decision making, etc. (Thakkar 2004). These viewpoints clearly reflect the centralized nature of national water policy production in the central Government that was a result of key drivers in the policy process, such as the bureaucratic elite, economists, political interests, and external forces. In spite of the fact that the circulated draft was critiqued with recommendations by civil society, the final policy did not incorporate the measures advocated.

The role of institutions within the state

With the National Water Policy (2002) in place, the Ministry of Urban Development went ahead in making major changes to allow 100 percent foreign direct investment (FDI) in urban infrastructure projects. This included development of water supply sources, water distribution, billing, sewage reclamation and reuse, management of unaccounted-for water, manufacture of water supply equipment, and privatization of solid waste management systems. At present, the central Government offers special incentives for investments such as exemption from customs and excise duties on imported machinery and exemption from all taxes for the first five years of water and sewerage projects (Rajamani 2004). The Government has provided these fiscal incentives to encourage partnership with the private sector and to attract foreign investment in urban water supply and sanitation projects. It has further amended the municipal acts to enable ULBs to partner with the private sector and to improve governance and management (Rajamani 2004).

A series of reports, presentations, and bureaucratic trips to Washington DC followed, pursuing and advocating policies for PPPs in water supplies and sanitation. The agreement over the Millennium Development Goals (MDGs) in the United Nations and the Johannesburg goals at the World Summit for Sustainable Development in 2002 committed India to achieving 100 percent coverage in urban water supply by 2007, which meant approximately 43 million additional urban people (at the current level) had to be covered. In order to meet the MDGs and Johannesburg World Summit goals in 2015, India will need resources for supplying urban water to approximately 88.5 million people. To meet the goals in 2025, it will need to supply water to approximately 236.5 million people over and above the 2015 levels. Extrapolation shows estimated investments in water supplies would be to the tune of Rs. 96 billion by 2015 and 258 billion by 2025 (Planning Commission 2002a: 54).

These statistics constitute part of the Planning Commission's report titled *India Assessment 2002 – Water Supply and Sanitation*, funded by the WHO and UNICEF. Based on the numbers, the report argues:

> If India's aspirations for continued economic growth and improved social and environmental conditions are to be met, fundamental changes in how water is allocated, planned and managed must occur. The currently ongoing reform process in Rural Water Supply and Sanitation and Urban Water Supply and Sanitation and New National Water and Health Policies are

important steps in the right directions. These should be sustained, and where necessary, augmented by further reform measures.

(Planning Commission 2002a: 11)

The report mentions that the supply-side approach to water has resulted in major economic, social, and environmental costs and emphasizes a demand-management policy.

Highlighting the policy objectives of the urban water supplies sector, including universal coverage, adequacy and regularity of water supply, and avoidance of excessive withdrawal leading to depletion, the report details the urban water problems that relate to cities in India. Poor quality of transmission and distribution networks, physical losses ranging from 25 to 50 percent, low pressures leading to back siphoning resulting in contamination, and water availability ranging from two to eight hours a day are issues which are largely real. However, what it recommends as policy strategies includes:

1 Decentralization.
2 Corporatization and commercialization of existing institutions.
3 Enhancement of technical and managerial capabilities.
4 Unbundling or rebundling of functions of ULBs.
5 Institutional restructuring.
6 Changing the role of Government from provider to regulator and facilitator.
7 Appropriate forms of private participation and PPS such as service contracts, leases and concessions, Build-Own-Operate (BOO) and Build-Operate-Transfer (BOT) project financing, etc. to be facilitated.
8 Benchmarking performance.
9 Evolution of a sound sector policy.
10 Water pricing based on volumetrics.
11 Transition from state monopolies to competition (Planning Commission 2002a: 56).

A consensus had emerged among the political and bureaucratic elite about the need for PPPs. A working group with representations from select ministries and the Planning Commission was first set up in the Prime Minister's office in January 2002. This group brought out a concept paper on PPPs in June 2003. The Committee of Secretaries of various ministries under the chairmanship of the Cabinet Secretary constituted a subgroup on infrastructure PPP under the chairmanship of the Secretary of the Planning Commission on 9 September

2003. The Planning Commission brought out its *Report of the PPP Sub-Group on Social Sector: Public Private Partnership* (2004b) in November of the following year. The report concedes that PPP is a business model but maintains that it should be introduced in different sectors with adequate understanding (Planning Commission 2004: i). The report defines PPPs to encompass all nongovernmental agencies such as the corporate sector, voluntary organizations, self-help groups, partnership firms, individuals, and community based organizations. PPPs would subsume all the objectives of the service being provided earlier by the Government, and are not intended to compromise on them. "Essentially, the shift in emphasis is from delivering services directly to service management and coordination" (Planning Commission 2004: 4). The report also emphasizes that PPPs lead to improvement in both "efficiency" and "effectiveness" in services (ibid.: 6).

The role of businesses

Apart from policy documents and reports, a major development was the involvement and the support of the Indian business industry that also moved into the water sector. At the Indian Economic Summit in New Delhi on 27–9 November 2005, the Indian Business Alliance on Water (IBAW) was launched with the support and partnership of the Confederation of Indian Industries (CII), the United Nations Development Program (UNDP), the United States Agency for International Development (USAID), the World Economic Forum (WEF), and the Prem Durai Exports-Switcher. The alliance is intended to facilitate the development of PPPs in water projects, broaden business sector engagement in commercial water projects, and promote corporate best practices in water. CII with the collaboration of the WEF Water Initiative hosted the water summit that facilitated PPPs in water and watershed management with the aim of bringing the latest trends, technologies, and best practices to Indian industry (26–7 November 2005) (CII Water Summit, Press Release, 2005 Delhi). In the words of Richard Samans (2005), Managing Director of the WEF:

> India is facing significant challenges regarding water access and quality, and the business community can be an important part of the solution by improving water management efficiency and working in closer partnerships with communities and municipalities. The Indian Business Alliance with communities has the potential to make contributions in this respect and the WEF is pleased to support it.

The assessment reports of the Government of India, the IBAW initiative, and the policies of international financial institutions like the World Bank and the Asian Development Bank (ADB) reiterate their commitment to private sector participation in water resources. The World Bank Water Resources Sector Strategy (WRSS) (2003) claims that water utility reform usually means substantial benefits for the poor and makes the water sector attractive to private investors. The ADB's Water Policy, approved in 2001, seeks to promote water as a socially vital economic good that needs careful management to sustain equitable economic growth and to reduce poverty (ADB 2001). The most recent addition to the internal reform measures advocated by the various government departments and the Planning Commission of India is the World Bank's report *India's Water Economy: Bracing for a Turbulent Future* (November 2005). The report clearly states that:

> India faces a turbulent water future. Unless water management practices are changed – and changed soon – India will face a severe water crisis within the next two decades and will have neither the cash to build new infrastructure nor the water needed by its growing economy and rising population.
>
> (Briscoe 2005: 4)

Its biases can be inferred from the following section:

> The state needs to surrender these tasks which it does not need to perform, and to develop the capacity to do many things which only the state can do. Competition needs to be introduced in the provision of basic public water services, bringing in cooperatives and the private sector. The state can then focus on financing public goods such as flood control and sewage treatment and play the role of a regulator to balance the interests of the users.
>
> (ibid.: 54)

Vigorously advocating the entry of private players into the water supply chain in India, the report argues that the presence of private players is essential, as there existed no civic institution in the country that could provide water supply 24 hours a day. The entry of private players would improve the quality of services of local bodies, exactly as it has compelled state owned enterprises in other fields to improve the quality of their own products and services (Ravindran 2005).

While conceding that the report is correct that investments in water infrastructures in India reflect a "Build-Neglect-Rebuild philosophy" (Mohanty 1995), it may be prudent to remember that the document was prepared by an agency that has a stake in water infrastructure. The World Bank's neoclassical thinking reflects a shift from Keynesian principles based on three major trends in infrastructure development: innovation in technology; the need for a shift from direct government provision of services to private sector participation; and increased concern for social and environmental sustainability. These are largely responsible for the heightened interest in water supply provision (World Bank 1994). Commercial principles, competition, and private sector participation, says the World Bank, will solve two problems in one: the state will have better maintained infrastructures and a more extended network could be created reaching more people (Finger and Allouche 2002: 71). In short, the World Bank believes that to promote environmentally sound and economically sustainable development, water resources have to be managed rather than developed. Such management is possible only if there is a paradigm shift from water as a public good to water as an economic good. In the process, the World Bank successfully transformed a "potentially explosive political question about rights, entitlements, how one should live, and who should decide, into technical questions of efficiency and sustainability" (Li 2002: 1).

The terms "decentralization," "unbundling," "management," "technology, "economic efficiency," and "sustainability" recur consistently in almost every report and also signify an apparent link between these reports and policy documents and the ideas and discourse of the elite and external support agencies. A careful reading of the various policy documents reveals a fairly coherent and interconnected set of ideas – ideas that seem to transcend national political boundaries.

1   India is facing a serious water crisis that needs to be urgently managed within a historic timeframe.
2   Water policy reform is essential for development, economic growth, good governance and access to water for the poor who suffer the most.
3   Poorly designed fiscal policies and governance constraints have further lessened access to urban water supplies by the people. In other words, the real flaw lies with inefficient and politicized Governments that treat water as a free public good.
4   The failure to charge people the use cost to reflect the true cost of water has inculcated a culture of wastefulness leading to crisis and scarcity of water.
5   Consequently, ensuring universal coverage and regularity of water supplies in a developing country like India not only requires economic instruments and private sector participation but also devolution of administrative responsibilities (decentralization) and PPPs, while the processes of privatization of resources and globalization of capital continue at an unprecedented scale.

The Planning Commission, the Ministry of Water Resources, and the Ministry of Urban Development all endorse the increased participation of the private sector. All these documents focus on a policy design that recommends private sector participation, technological innovation, economic and institutional reform for universal coverage, and efficiency and sustainability of water resources. These documents also expose how the concept of "economic efficiency" has come to dominate developmentalist thinking in India and the manner in which policies have been formulated to achieve those ends.

The analysis reveals that this process was facilitated gradually, during which time a key challenge for the Government was balancing the domestic interests of the constituencies with the conditionalities of the external agencies. The structural linkages between bureaucrats, politicians, external forces, and well-qualified economists, some of them with long working experience in, and a thorough grasp of the structural power of, international financial institutions like the IMF and the World Bank, created the opportunities to push forward these reforms. This knowledge gained ascendancy in the national agenda through the authority exercised by the key circle of policy makers with strong political support and the political and economic imperatives managed by the World Bank. Under a neoliberal vision of poverty alleviation and ecological sustainability, a consensus emerged in the corridors of the Government that this new political rationale of development, in which private sector participation in water issues is essential, is the approach best capable of serving society in India.

This national agenda for water policy reform has been under scrutiny by various groups within civil society. But for a Government steeped in the ideology of economic growth through liberalization, privatization, globalization, and the rollback of the state, and a bureaucracy that has enjoyed the fruits of centralized control since the colonial period, it would be difficult to produce a radical vision for an environmentally sound and socially just development for the country (Lele and Menon 2004). On a national level, the policies are in place, but these reports and policy documents are subject to implementation by the individual states. What one can establish from this overview is the rapid pace of change, the increasing visibility of the politics of policy production, and corresponding changes in the role of the center that allows private investment, particularly foreign, to participate in a strategic sector like water. The impact of liberalization in general and water reform policies in particular created its own set of subnational dynamics for individual states to adopt a water policy based on national guidelines.

Liberalization, subnational dynamics and the water sector

The implementation of the first generation of reforms under the political support of the leaders and the structural adjustment programs of the IMF and the World Bank led to considerable tension between the center and the individual states. The budget reflected the center's concern about fiscal deficits, with major cuts in social spending in such centrally sponsored schemes as irrigation and water supplies (Seetaprabhu 2001). Facing a financial crunch on transfers from the center, states found it increasingly difficult to implement election promises made on such populist measures (Shastri 2001: 16). The principle by which states are expected to pay their share of bills to the public sector had begun to evolve during Indira Gandhi's time and continued to strengthen under the regimes of Rajiv Gandhi and Narsimha Rao. The crisis in Rao's rule compelled state Governments to manage their own dues or have the central Government deduct it from their plan grants.[4] This had an adverse bearing on the economies of various states that constitute the Union of India:

1 The move precipitated recessionary conditions, lowering the levels of economic activity in the states by reducing revenues generated through taxes imposed by the state Governments – sales taxes, for example. The reduction in the state Governments' income adversely affected its capacity to provide for the welfare of its people.

2 The overall reduction in the expenditures by the central Government meant that the central transfers to the state Governments (plan transfers) were sharply reduced. This compounded the plight of state Governments already suffering from reduced revenues.

3 The central Government, in order to bolster its own revenues, also began to charge extremely high rates of interest on the loans they advanced for development of the states and then advanced the value of the interest to the states at even higher rates than the original loans.

4 Reduction of direct taxes affected the states' share of central taxes because much of the central tax revenue was shared with the states. A reduction in the central direct tax revenue quantum meant a corresponding decline in the states' share of the central tax revenue.

It was natural under the circumstances that many states began to take initiatives for fiscal reform. With the central Government allowing private sector investment in infrastructure, a number of states started to negotiate agreements with the private sector in power, roads, water, and other development projects. Reforms shrouded in politics were pursued in a piecemeal manner at the state level. With the share of funds curtailed by the central Government policies, many states started to seek alternative sources in the form of loans and grants from international agencies. An overall climate was in place that allowed these agencies and multinational corporations (MNCs) to pressure states to liberalize, privatize, and globalize (LPG).

Bilateral agencies like the UK's Department for International Development (DFID), the USAID, Germany's Federal Ministry for Economic Cooperation and Development (BMZ), and Japan's Overseas Economic Cooperation Fund (OECF) have also moved into various projects and policy issues relating to drinking water, sanitation, irrigation, and the like at the national and subnational levels. Couching their offers in the language of technological assistance and transfer, these agencies have been making inroads into the policy space to generate markets for the entry of MNCs under PPPs. These agencies have worked toward providing market access for their domestic corporations to carry out studies to construct and operate water treatment plants (Shiva 2002). Most also provide consultants to assess the "efficiency" and "cost effectiveness" of various schemes.

The World Bank and the ADB also moved into the individual states. The loans mostly came with structural adjustment programs that entailed privatizing certain sectors, cutting subsidies, downsizing staff, and meeting certain fiscal conditionalities. The World Bank noted that the move to focus large scale integrated investment packages on the few states willing to undertake public sector expenditure reform gave it much more leverage than it had before (Pitman 2002). The first state to move in this direction was Andhra Pradesh. However, an analysis of the economic restructuring program of Andhra Pradesh shows that the greater leverage of the bank led to the opening up of a policy space that moved in the direction of the commercialization of socialization and infrastructure, with other states following suit.

The states proceeded to incorporate changes in the state water provisions as a result of the developments that occurred on the national scene and to enhance state capacities to sustain these programs. These developments included the heightened role of international agencies, the financial problems of the states due to cuts in social spending by the center, and the political interests of politicians to demonstrate the progressiveness of the state in developmental and welfare projects. Maharashtra, Karnataka, Andhra Pradesh, Rajasthan, Tamilnadu, and Madhya Pradesh added the clause of private sector participation in some form to their water policy documents.

Two projects initiated at the outset reflect the nature of the trends in water policy reform. The agreement for the Sangli-Miraj-Kupwad project, which was financed by the USAID and Infrastructure Leasing and Financial Services (ILFS), had the following conditions with respect to private sector providers. The ILFS would answer no queries raised by the public on water issues and would have all rights over the water supplies. It would be the duty of the municipal corporation to generate the revenue demanded by the ILFS and other private US entities. The ILFS and the US private entities would have rights over the property and stocks of municipal corporation (Barve 2002). A similar project taken up by the state Government of Tamilnadu was the Tirupur Area Development project, aimed at initiating private sector participation with US based companies like Bechtel and to persuade the state machinery to experiment with these concepts of privatization (RFSTE 2005; Painter 1999).

In 2001, the ADB announced its Water Policy, which focused on expanding water services delivery through autonomous and accountable service providers, private sector participation, and PPPs. The policy aimed to reallocate water through "markets of transferable water rights" and stressed that the Government needs to modify its role from "one of service provider to regulator" (ADB 2001). Under its broad umbrella of "poverty reduction in urban areas," the ADB moved into the states of Karnataka, Rajasthan, Kerala, Madhya Pradesh, and Sikkim through its country assistance plans that intensified private sector participation.

The World Bank and the other donors have supported their justification of private sector participation by "research studies" and "surveys" through consultants commissioned by them, such as PricewaterhouseCoopers and GKW. These highly paid consultants generate the numbers and statistics to strengthen data in order to legitimize the approach of these agencies. These reports reaffirm water scarcity, inefficiency, cost recovery and effectiveness, and effective water solutions at the local level in the state consultancy projects.

There is no doubt that "price" is an issue that needs to be addressed in the Indian context, and there is truth in the fact that subsidies actually do not reach the poor, but the solution offered in terms of private being effective and public being ineffective does not reflect the Indian understanding of sociopolitical realities. However, given the crisis facing the states and the ambition of politicians to showcase their progressiveness, combined with their interests in the power exerted by international agencies, a shift occurred in the water policy of the states, most of which went ahead with PPPs. The process continues at an unprecedented pace. Already some 30 cities in Maharashtra, Karnataka, Andhra Pradesh, and Rajasthan have tendered their respective municipal water to a handful of MNCs, even though experience from cities of other developing countries shows that the privatized and commercialized supply of water often deprives the poorer and marginalized sector of their basic right to water (RFSTE 2005: 11).

In this context, the following chapter situates Delhi within the water reform process. The process of reform in the water sector in Delhi needs to be understood in the political and sociocultural milieu of the region. Chapter 5 introduces Delhi's geographical, topographical, and hydrological background. In addition it presents the Delhi Jal Board's policy vision under which the first step toward reform began with the inauguration of Sonia Vihar for the 24/7 project.

Notes

1 See D. D. Kosambi (1965), D. N. Jha (1998), Romila Thapar (1966), A. L. Basham (1967), Shareen Ratnagar (2001), and Ranbir Chakarvarti (1998).
2 The East India Company was granted a monopoly license to trade with India in 1600 which continued until 1813 when other private capitalists started protesting over the monopoly and wanted the opening up of free trade in India. It can be said that colonial rule in India had its origins in the East India Company.
3 The period 1995–99 was a period of great uncertainty at India's center.
4 The center–state relations in India have a history of central control over revenue. The center distributes moneys to the individual states in the form of grants for development and the social sector. The states' lack of fiscal autonomy and their fiscal dependence on the center has placed the states in debt to the center. Occasional waivers have added to further fiscal laxity on the part of the states (Weiner 1999). Liberalization adversely amplified these effects.

5   Situating Delhi in the water reform project

This chapter opens with a general introduction to, and overview of, Delhi, with an emphasis on the city's water resources and urban water delivery mechanisms, before using the city as a test case for the water reform project. The information about the region's current water scenario, and proposed reforms to meet the city needs, provide the necessary background for competing policy discourses on the water sector reform process under the Government of Delhi. The chapter is divided into five main sections: the first discusses the rise of Delhi as a global city; the second describes the geography, topography and population of the region; the third highlights the availability of water resources, treatment and distribution in the city; the fourth examines the current water scenario in Delhi; and the fifth describes the Government's proposed solutions to meet Delhi's water needs, which were contested by nongovernmental organizations (NGOs) and the people of Delhi.

Delhi – the rise of a global city

The city of Delhi has historically been one of the most important cities of India. Delhi hosted the British seat of governance when the capital moved from Calcutta in 1911. Today, Delhi hosts both the Government of India and the State Government of Delhi.[1]

The Delhi of the 1990s may be described as a post-colonial city with a first world desire (Nigam 2001).[2] Political leaders are also working to transform Delhi into a global city. In part, this desire of the politicians and planners to make Delhi into another global metropolis can be ascribed to the rapidly emerging "new global order."

The *Economic Survey of Delhi 2001–2002* (Planning Department 2002: Ch. 9) stated that:

in order to make the National Capital Territory of Delhi a world class cyber state, the Delhi Government has planned to set up a state-of-the-art 'Hi-tech City for Information Technology' on about 100 acres of land with the best communication links, uninterrupted and clean power supply with backup power generation, and other advanced technical facilities. A World Trade Centre is being set up to provide instant information flow and interaction among various players in international trade and commerce. With a view to establishing and developing industry–university research and development linkages, a 'Bio-Technology Park' is also being developed in collaboration with Delhi University.

The Chief Minister of Delhi, Sheila Dixit, has led a Congress Party Government since 1998 and describes her Government's vision as one of making Delhi a "world class city" (Sethi 2005; *The Tribune* 14 November 2008) with modern infrastructure catering to the needs of its citizens. The Congress Party President, Sonia Gandhi, added that Delhi's development sent a message to the entire country and hence technology development should remain the agenda of Government in Delhi (*The Hindu* 20 November 2004). The city over the past 30 years, has developed 31,183 kilometers of roads, 34 flyovers and 65.1 kilometers of mass rapid transit system and is looking for more infrastructural development before it hosts the Commonwealth Games in 2010 (see Figure 5.1 for the location of Delhi in India).[3]

The Delhi Urban Environment and Infrastructure Improvement Project (DUEIIP 2001) report, submitted to the House (the legislative assembly of the National Capital Territory (NCT) of Delhi) in March 2001, lays down a strategic framework for the development of Delhi beyond 2000. The report outlines the strategies and the policy changes required to make Delhi a more environmentally sustainable and livable city by 2021, keeping in view its potential growth and population. The Government of Delhi, the Ministry of Environment and Forests, and the Government of India sponsored the report, which had Japanese funding through the World Bank.

Chapter 9 of the report describes in detail the policy objectives and long-term goals in water management of the NCT of Delhi. The Government of Delhi's long-term goals as recognized in the DUEIIP report are to:

1   Serve all areas of the NCT including planned growth areas with water supply.
2   Reduce environmental impacts of poor sanitation.

3 Ensure wastewater discharge to the Yamuna River is within prescribed quantity and quality limits.
4 Reduce water demand and improve service delivery through increasing efficiency and commercial autonomy.

The overarching long-term goal is to provide all NCT citizens with equal access to an adequate quantity of potable water within the available natural water resources (DUEIIP 2001: 38). The Government of Delhi argued that the preferred strategy

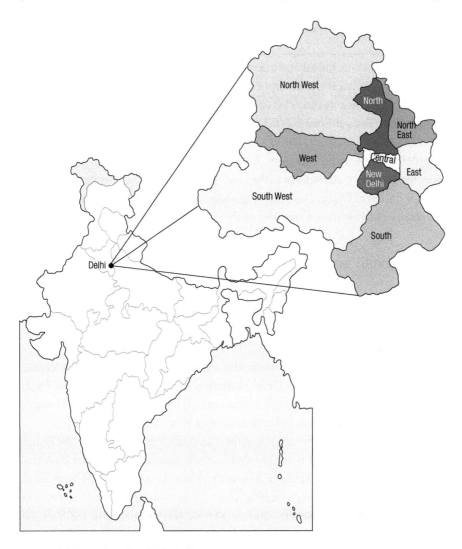

Figure 5.1 Location of Delhi in India.

Source: Based on http://www.delhilive.com/system/files/images/LocationofDelhi2.preview.JPG

to achieve these objectives required establishment of a phased program of full service coverage and high quality provisions. Its vision document discussed in DUEIIP (2001) sets out a time frame up to 2015 in three five-year phases, with the first phase beginning in 2001.

In order to accomplish the objective of the DUEIIP (2001) report, the Government of Delhi promised the following reforms:

1 Improve efficiency of service delivery systems.
2 Enhance technical capacity and capacity building.
3 Comprehensive training of the officers at national and international institutions.
4 Develop key technical and managerial competencies.
5 Harness the latest technologies.
6 Outsource various activities to improve quality and cost efficiency (Finance Minister Budget Speech 2005–6).

In order to understand therefore, how the Government of Delhi proposed to meet the goals of water reform in the city, one first needs to understand the socio cultural milieu of Delhi. The following section provides some background to the geography and population scenario of the region.

The geography and population of Delhi

Geographical area

Delhi covers an area of 1,486 sq. km., of which the urban area accounts for 525 sq. km. (Planning Department 2008: 2). Out of the total urban area, the area under the jurisdiction of the Municipal Corporation of Delhi (MCD) accounts for 94 percent, while the New Delhi Municipal Council (NDMC) and the Delhi Cantonment Board (DCB) cover 3 percent each (DJB 2004). The MCD is one of the largest municipal bodies in India according to the population size.

Delhi sprawls over the west bank of the Yamuna River and is one of the fastest growing cities in India. It is surrounded on three sides by the state of Haryana, and to the east by Uttar Pradesh, across the Yamuna, which flows from north to south. A hard rocky ridge running from the southern border of the NCT in the southwest in a northeasterly direction to the western banks of the Yamuna near the Wazirabad Barrage forms the main watershed in the NCT (Planning Department 2004: 1–6). The topography creates a drainage system that carries rain and storm water from the higher elevations of the west to the Yamuna. The eastern, low lying side was originally part of the flood plain of the river and considered inhabitable. Today, however, this eastern wing, also known as the Trans Yamuna Area, houses about 23 percent of the total population of Delhi (DUEIIP 2001).

While Delhi is located on the west bank of the Yamuna, its proximity to the Ganges is important. The city is effectively at a juncture of the Ganges River Valley and the Indus River Valley where the Deccan Plateau and Thar Desert come closest to the Himalayan Mountains, creating the corridor where Delhi is located.[4]

Population: vastly outgrowing capacity

Due to explosive growth patterns, in part triggered by political refugees but also fueled by unique employment and investment opportunities, the Delhi water supply as currently managed is inadequate for present needs and promises to be even more so in the near and continuing future.

Delhi began as a small city with a population of 410,000 in 1911 and reached a population of 920,000 in 1941 with a decadal growth averaging approximately 30 percent. In the 1960s, the population of three million was supportable by its natural water resource base. In 2004 the city had a population of around 150 lakhs (one lakh = one hundred thousand) (DJB 2004: 1), attracting more than four lakhs of new entrants each year (DUEIIP 2001). The result has been a concentric expansion of the city and a high increase in the disaffected population (almost 50 percent of the total), leading to a breakdown in essential services, health, water sanitation, and pollution. By 2021, the population will be over 230 lakhs (23 million) with severe water and land shortfalls (DUEIIP 2001; DJB 2004: 1). Currently it is the third largest city in terms of population, exceeded only by Mumbai (Bombay) and Kolkatta (Calcutta). Explosive population growth leading to a breakdown in water services, as well as the vision of Delhi as a global city motivated the Government of Delhi to invest and reform the infrastructure of the city, of which water is a major component.

Sources of Delhi's water supply

A brief backgrounding of Delhi's current water sector will help to better analyze the process of policy design and the implementation of its objectives and long-term goals in water supply.

Water availability, treatment and supply in Delhi

According to the CSE (2003), Delhi receives water from 3 sources:

1 **Surface water.** Eighty-six percent of Delhi's total water supply comes from surface water, namely the Yamuna River, which equals 4.6 percent of this resource through interstate agreements.
2 **Subsurface water.** Ranney wells and tubewells. This source, which is supplied by approximately 611.8 mm of rainfall in 27 rainy days, and unutilized rainwater runoff, totals 193 million cubic meters (MCM).
3 **Graduated resources.** These are estimated at 292 MCM, however current withdrawal equals 312 MCM. Salinity and over exploitation has contributed to depletion of the resource and drastically affected the availability of water

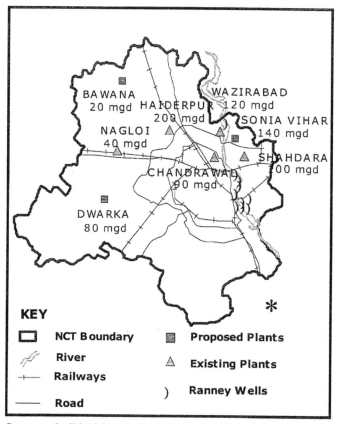

Source: Delhi 1999 – A Fact Sheet, NCRPB

Figure 5.2 Existing and proposed water treatment plants.
Source: Based on DUEIIP (2001: 41).

in different parts of the city. However, according to a report released by the Central Ground Water Board (CGWB 1999), Delhi's groundwater level has gone down by about eight meters in the last 20 years, at the rate of about a foot a year.

Apart from groundwater, Delhi gets water from the Ganga Canal, the Western Yamuna Canal and the Bhakra Canal (CSE 2003). In total, water production by the Delhi Jal Board (DJB) – see Table 5.1 – is currently estimated at 650 million gallons per day (MGD), assuming that the main water treatment plants are producing at rated capacities.

Most of the surface water that the DJB controls is treated at five plants: Chandrawal, Wazirabad, Haiderpur, Bhagirathi, and Nangloi, which have a combined treatment capacity of 640 MGD (2911 million liters per day) of water. Delhi proposes to augment water supply by building three more plants (Figure 5.2).

The Government of Delhi is responsible for both urban and rural water supplies. The DJB was constituted on 6 April 1998 through an Act of the Delhi Legislative Assembly which incorporated the previous Delhi Water Supply and Sewage Disposal Undertaking, is responsible for the production, treatment and distribution of potable water after treating raw water from various sources like the Yamuna River, the Bhakhra Storage, the Upper Ganga Canal, and groundwater. However, under the current water scenario, the DJB is unable to meet the needs of the city.

The current scenario in Delhi's water services

Water demand exceeding supply

Since the pressure on water resources has been increasing proportionately with the increase in population, rapid urbanization, unplanned and uncontrolled growth, shanties, and migrant population, the water supply infrastructure has come under severe strain. The present water demand is 850 MGD and is expected to be around 1,380 MGD in 2021 (DUEIIP 2001; DJB 2004: 1). In 2004 Delhi's water supply capacity was 650 MGD and depended on the various sources of raw water (DJB 2004).[5] Table 5.1 shows the various sources of available raw water, and the installed water supply capacity of the water treatment plants that are functioning under the DJB.

While the DJB had the capacity to supply 650 MGD of water in 2004, with 20 percent transit losses, it actually supplied around 520 MGD water, which led to a shortage of almost 330 MGD in the city (Gupta 2003). There exists

Table 5.1 Delhi's water supply capacity in millions of gallons per day (MGD)

	Sources of raw water	Name of plant	Installed capacity
1	Yamuna River	Chandrawal I and II	90 MGD
2	Yamuna River	Wazirabad I, II and III	120 MGD
3	Bhakra Storage	Haiderpur I	100 MGD
4	Yamuna River	Haiderpur II	100 MGD
5	Bhakra Storage	Nangloi	40 MGD
6	Upper Ganga Canal	Bhagirathi	100 MGD
7	Subsurface Water	Ranney wells/Tube wells	100 MGD
Total			**650 MGD**

Source: DJB (2004: 1)

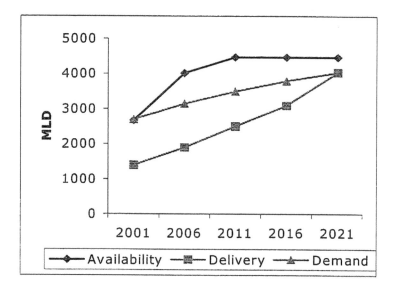

Figure 5.3 Demand, supply and delivery projections in millions of liters per day (MLD).
Source: Based on DUEIIP (2001: 41).

therefore a demand supply gap in the water provisions of Delhi. Future projections of the DUEIIP (2001) report show an increasing demand and supply gap (see Figure 5.3) that the Government of Delhi feels can be bridged by efficient delivery services.

Irregular and intermittent water supply in Delhi

Currently, water supply does not reach all parts of the city, and water in Delhi runs only 4–6 hours a day. Intermittent supply, seasonal disruptions, insufficient and

irregular pressures and unreliability of supply are some of the major issues facing households in Delhi (Ruet and Zerah 2001; Ruet 2002; CSE 2003). In many
areas of Delhi, women have to get up early in the morning to collect water for
household needs because water in those areas is limited to less than 2-hour-long
morning and evening shifts (Canepa 2004).

Delhi water distribution is inequitable

Although water availability is estimated at 240 liters per capita per day (LPCD),
the highest in the country, these figures do not indicate an adequate supply to
every resident (CSE 2003: 3). This inadequacy is mainly on account of inequitable distribution of water and loss of water through leaking pipes, leading to
abundance in some areas and acute shortages in the others. "An average room in
a five star hotel in Delhi consumes 1,600 liters of water every day. VIP residences
consume over 30,000 liters per day. The Prime Minister's house at 1 Race Course
Road accounts for 73,300 liters of water per day and the Presidential residence
consumes about 67,000 liters per day but 78 percent of Delhi's citizens, who live
in substandard settlements, struggle to collect or buy 30–90 liters of water per
capita per day" (Singh 2005). The NDMC and DCB areas get an average supply of above 450 LPCD, while areas in Narela and Mehrauli zones get less than
35 LPCD (see Figure 5.4, Delhi Fact Sheet 1999, cited in DUEIIP 2001).

Percentage of population with inadequate water supply

While the treated water in Delhi is adequate for supply to consumers, 10 percent
of the Delhi population have no piped water supply and 30 percent have grossly
inadequate water supply. Even planned areas of the MCD, with house connections, have a shortfall of 42 percent (see Table 5.2).

Water shortages due to nonrevenue water[6] in Delhi

Delhi's water losses amount to up to 50 percent due to leakage, unauthorized and
unbilled consumptions. The DJB recovers only 40 percent of its operating costs
through billing (Sharma 2005). The apparent losses are 2 percent and real losses
40 percent. In total 40 percent of water is lost due to leakage (see Table 5.3). This
leakage of water as well as the inadequate maintenance of the water treatment
plants creates shortages and supply. DJB's losses due to lack of cost recovery are
currently in crores[7] of rupees and the institution itself is marred by inefficiency,
corruption and unaccountability (Sharma 2005).

Table 5.2 Percentage of population with inadequate water supply in Delhi

	Type of settlement	Population in lakhs (100,000s)	Demand in millions of liters per day (MLD)	Supply in MLD	Shortfall/ excess
1	Jhuggi Jhopri[1] cluster, designated slum area and unauthorized colony (I)	13.96	59.33	No piped supply	(–) 100%
2	Jhuggi Jhopri cluster, designated slum area and unauthorized colony (II)	40.80	173.40	20.43	(–) 88%
3	Planned area	75.50	1,698.75	990	(–) 42%

Source: DUEIIP (2001).

Note
1 A Jhuggi Jhopri is a small, roughly built house or shelter usually made of mud, wood or metal with a thatch or tin sheet roof covering.

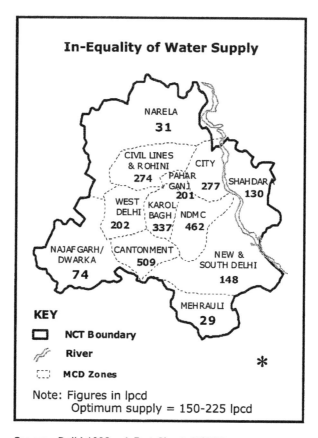

Source: Delhi 1999 – A Fact Sheet NCRPB

Figure 5.4 Inequality in Delhi water supply.

Table 5.3 Nonrevenue water in Delhi

System input volume	Authorized consumption 58%	Billed authorized consumption	Billed metered consumption (including water exported in bulk)	13%	Revenue water 50%
		Unbilled authorized consumption	Unbilled unmetered consumption	8%	
			Unbilled metered consumption	0%	
	Water losses 42%	Apparent losses 2%	Unauthorized consumption	2%	
			Metering inaccuracies	0%	
		Real losses	Leakage and overflow at utility's storage tank	0%	
			Leakages on distribution mains and service connections up to point of customer metering	24%	

Source: Planning Department (2008: 164).

Private operators to meet water shortages

Due to the shortage of water in Delhi, private water tankers supply water in areas where the DJB water does not reach the people. According to the Water Tankers Association there are 250 private water suppliers supplying water through tankers in Delhi. The water tankers supply untreated water and they do not take responsibility for its quality. These suppliers get water by drilling bore wells and tube wells. Private water tankers have come into existence precisely due to the failure of the Government of Delhi to meet the water demand of the city. These tankers supply water on behalf of the DJB to several slum and rural areas of Delhi, colonies under the jurisdiction of the Delhi Development Authority (DDA), regularized colonies, and upcoming societies (Llorente and Zerah 2003). They also supply water on behalf of the DJB to government hospitals, the central jail, congested colonies, and Metro projects. In summer, these tankers supply water to individuals and the DJB hires 200–400 water tankers from private parties during periods of shortages. Apart from tankers that operate on behalf of the DJB there are also private tankers that operate in water scarce areas charging high prices of the consumers. The water sold is often unfit for human consumption (Daga 2003: 176).

The above issues facing Delhi's water supply system reveal that it is mismanaged and currently inadequate to meet the present needs of the population and will be even more so in the future, unless the Government takes strong measures to improve and redesign water supply provisions.

Proposed solutions to the existing water problems

In 2001 it was proposed to launch the reform process immediately, with the objective of achieving visible and replicable improvements in various aspects of the water supply and sewerage (WSS) services in Delhi. The Government's plan involved three steps:

1 Augmentation
2 Treatment
3 Distribution

The Government proposed that in this process of improving water supply and services, supply of raw water, treatment, bulk distribution, and retail distribution would be handled by different parties in contrast to the post-independence models of a monopoly, which saw authorities like the DJB handling of all these functions.[8]

Augmentation

The DJB decided to augment its water supply capacity by commissioning the following new projects: the Bawana Plant, which would have 20 MGD water treatment capacity, and the Sonia Vihar Plant which would treat 140 MGD. Other steps included optimizing the capacity of the existing water treatment plants from 19 MGD to 40 MGD (Table 5.4). This would help meet the demand of Delhi's then current water needs of 850 MGD.

Raw water for the Sonia Vihar Plant

The Government's first step towards a comprehensive reform policy was the augmentation of 300 cusecs of raw water from the Tehri Dam (see Figure 5.6 and 5.7) to Delhi. To carry this raw water, a conduit of 3,250 mm in diameter was laid from Murad Nagar to Sonia Vihar (see Figure 5.5) through Uttar Pradesh Jal Nigam at an estimated cost of Rs 110.00 crores (Shiva and Jalees 2003). This water was to be transferred to the water treatment plant at Sonia Vihar in Delhi which is now working under capacity due to lack of available raw water since November 2006 (*The Tribune* 14 April 2007).[9]

Table 5.4 Proposals for augmentation of water supply

	Name of the plant	Existing capacity on 31 March 2004 in MGD	Existing capacity on 31 March 2008 in MGD
1	Chandrawal Water Houses I and II	90	90
2	Wazirabad I, II and III	120	120
3	Haiderpur	200	200
4	North Shahadra	100	100
5	Bawana		20
6	Nangloi	40	40
7	Sonia Vihar		140
8	Ranney wells/tube wells	81	87
9	Optimization of water treatment plants	19	40
10	Recycling of wastewater at Chandrawal, Bhagirathi, Haiderpur and Wazirabad	10	45
11	Iron removal at Okhla	–	12
12	Additional 20 MGD at Okhla	–	20
	Total	**650**	**914**

Source: Planning Department (2008: 162).

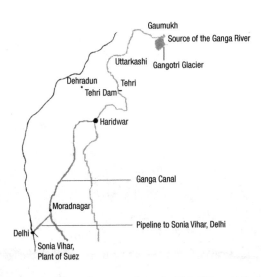

Figure 5.5 Map of water augmentation route from the Tehri Dam to the Sonia Vihar Plant, Delhi.

Source: Based on RFSTE (2005: 64).

Figure 5.6 Augmentation of water from the Ganga River near Old Tehri Town.

Figure 5.7 The Tehri Dam under construction.

Figure 5.8 The Ganga Canal at Hardwar.

Steps towards augmentation of water were initially delayed because of protests by farmers in that region who argued that this diversion of water from the Upper Ganga Canal (see Figure 5.8) to the French multinational Suez would be at a huge cost to their agricultural production.

The controversy started with the awarding of contracts to private firms. While the National Building Construction Corporation (NBCC) was given the contract for building the conduit in Uttar Pradesh (see Figures 5.9, 5.10), when work started on the Rs 880-crore project in November 2000, farmer groups in Uttar Pradesh started protesting.

The Dehat Morcha, a farmers' organization, staged protests and rallies. Their reason for the protest was that the diversion of water to Delhi would adversely affect their crop yields in lean seasons. The protests delayed the work on the conduit. Completion was further delayed when a part of the conduit, which was passing through a marshy area, gave way and water flooded the area (*Times of India* 17 June 2005). The raw water to be supplied from the Tehri Dam also ran into difficulties due problems in the tunnel of the dam. However, in spite of the initial delays, Delhi has now started receiving some water from its neighboring state Uttar Pradesh. This state supplies only 60 MGD raw water per day to the 140 MGD Sonia Vihar Plant (*The Tribune* 14 April 2007), which means that

Figure 5.9 NBCC pipes.

Figure 5.10 Pipeline carrying water to Delhi.

the plant is functioning under capacity, a fact that has major implications for the people and for the DJB, under the terms of the contract.[10] In 2007 a total of 720 MGD of water was being supplied to Delhi against the demand of 850 MGD (*The Tribune* 14 April 2007).

Treatment of water at the Sonia Vihar Plant

As part of the second step towards comprehensive water policy reform, the Sonia Vihar Plant was constructed to treat raw water and supply treated water at the main input zone for distribution to the citizens of South and East Delhi.[11] The plant at Sonia Vihar in Delhi (see Figure 5.11) was built at a cost of Rs. 1.8 billion. The plant was inaugurated for construction on 21 June 2002 (*The Tribune* 22 June 2002), and was designed to treat 635 million liters of Ganga water a day. Contractual parties were the DJB and the French company Ondeo Degremont

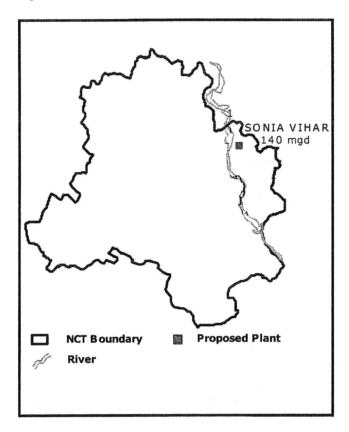

Figure 5.11 Location of the Sonia Vihar Plant in Delhi.
Source: Based on DUEIIP (2001: 41).

(a subsidiary of Suez Lyonnaise des Eaux—the world's biggest water giant)[12] (*The Hindu* 19 December 2003). The plant, originally intended to be complete by August 2003, was to provide Delhi with 140 MGD of treated water to meet its water requirements.

According to the DJB officials (Personal Interview, 4 December 2004), and government reports, the plant uses technology known as Pulsator Clarifiers and Aquazur V filters. "This unit operates under the principle of an up flow current through a suspended sludge blanket. Its distinctive characteristic is the utilization of a Pulsator technology. Pulsator clarifiers use a perfectly homogenous, stabilized sludge blanket for the removal of colloids, plankton, and algae" (CPCB 2004). This method is extremely efficient (95–9 percent) in the removal of algae compared to the conventional methods used by the DJB. "The *Aquazur V filters* use a single layer of homogenous sand and are suitable for high-rate filtration and an extended filtration cycle due to their in-depth clogging capacity" (CPCB 2004).

Sonia Vihar Plant disputed from the very beginning

Disputes closely followed the inauguration of the plant. First, some environmentalists and social activists challenged the building of the Tehri Dam on the Ganga River and the Sonia Vihar Plant in New Delhi. Questions were raised about Suez-Degremont's failure to pay any of the social, ecological or financial costs for the construction of Tehri Dam as well as the Sonia Vihar Plant while the DJB paid for all costs from infrastructure to supply of raw water, land and subsidized electricity to run the plant. At the same time Degremont was not responsible for transmission losses and revenue collection and was also assured the purchase of treated water and productivity incentives once the plant began operations (Kaur 2003). The objection of the protestors was that the corporation would get water free without paying for the social and environmental costs to those rural communities from whom the water would be taken (Shiva 2004; Dehat Morcha 2004).

Second, experts challenged the Government's justification that "state of the art technology" was not available in India[13] and that the technology chosen was the best avaiable. Third, several allegations were made that the tendering process was faulty and the contract negotiations and awards were nontransparent (Personal Interview of Jafri 2004; Shiva 2004; Sehgal 2007). The leader of the opposition parties and some ruling party members leveled allegations of corruption and irregularities in the allotment of the contract to Degremont (Mehdudia 2000). The Central Vigilance Commission of India instructed its technical examination committee to investigate why the contract, which was originally worth Rs. 295.75 crores ($70 million), had been awarded for almost Rs. 900 crores ($216 million) without a new tender process (Sethi 2005; Sehgal 2007; *The Hindu* 2005).

Lastly, people opposing the plant also questioned the performance based incentives for the company and penalties for nonperformance by the companies. If the DJB failed to supply the water to the company, a penalty of Rs. 50,000 per day would be paid to the company.[14] The contract imposed stiff penalties on the DJB and offered Ondeo Degremont easy incentives. The contract also provided for incentives for overproduction and energy savings (Sethi 2005). Critics thus argued that the provisions of the contract heavily favored the company over the DJB.

However, the strongest of the allegations was the fact that the entire process of reform, facilitated with the commissioning of the Sonia Vihar project, led to the commoditization of water in Delhi. The contract, which relied on water supplied from the Ganga, led to a nationwide movement against the privatization of water in Delhi, using the Ganga River as a symbol for water. The Water Liberation Campaign (WLC) began the struggle for their water rights but building water democracy meant building alliances with citizens' groups in Delhi and people's movements along the Ganga who helped frame the struggle against privatization. Millions signed petitions saying "Our Mother Ganga is not for sale." A Jal Swaraj Yatra (a water democracy journey) was organized in the years 2002, 2003, and 2004. Ganga Yatras were organized to rejuvenate the living culture of the sacred river. The Yatra in 2004 took place from 15–22 March, on World Water Day. A million people were contacted and 150,000 people signed a hundred-meter "river" of cloth to protest privatization (Shiva 2005; RFSTE 2006; *The Hindu* 19 December 2003). In a public resolution, the WLC demanded that the Government of Delhi withdraw its decision to set up a Regulatory Commission, make public all papers related to the privatization of the water sector and cancel all contracts with multinational corporations (MNCs) working with the DJB. Condemning the Government's proposal to levy user charges on extraction of ground water, the resolution demanded public–public partnerships (PPPs) instead of privatization (*The Hindu* 18 October 2005).

The DJB, however, announced that it held full ownership of the plant and claimed that the contract with Suez Degremont was aimed at allowing them to get the latest technology, as well as to learn how to use it during the ten year period of the operation of the plant by the company.

It was in the midst of all these allegations and objections between 2002 and 2004[15] that the DJB announced a proposal for a pilot project that would involve the distribution of treated water that came from the Sonia Vihar Plant into two zones of South Delhi as the final step in achieving its water vision as outlined in the DUEIIP (2001) report. Once successful, this pilot project would be replicated in the 21 distribution zones of Delhi. Thus, the Sonia Vihar Plant was not a standalone project, but a template for the future water production and distribution in the national capital. However, it was from this plant that Delhi's 24/7 water project was to become operational (this project was to be operational in November 2005. The 24/7 project was stalled on 23 November 2005 when the Government of Delhi wrote a letter to the Government of India stating that it wished to withdraw its loan application from the World Bank as a result of the year-long agitation against the Government of Delhi). The project was a vital part of the administration's attempt to make Delhi into a world class city, in spite of the protests and opposition that continued in Delhi throughout 2002–5.

Distribution of water supply in South Delhi (24/7)– zones South I and II

Assuming that the Sonia Vihar Plant would be functioning by August 2003, a pilot project was initated by the DJB that handed over the management of two DJB operating zones, South I and South II, to MNCs, as part of the Delhi Water Supply and Sewerage Sector Project that was funded by the World Bank.[16] A significant part of the 140 MGD of water processed by the Sonia Vihar Plant was to be allocated to South I and South II where this pilot project was under way. In accordance with the project requirements, the water distribution services of these two zones were to be handed over to private corporations on the condition that they supply water around the clock at a specified pressure for five years. After the trial period, the DJB would invite tenders for all 21 zones in Delhi, paving the way for dramatic water sector reforms in the national capital.

To accomplish this goal, under the directives of the World Bank, the DJB commissioned the *Delhi Water Supply and Sewerage Project Preparation Study.* The study, conducted by PricewaterhouseCoopers (PwC), assessed Delhi's water conditions in 2002 and focused on both technical and institutional aspects. The study gave a complete snapshot of the existing infrastructure and institutional arrangements and identified certain areas requiring critical intervention. These were organizational restructuring, corporate governance, the regulatory mechanism, financial sustainability and reliability of supply services to the entire city (PwC 2004). Three other consultants, GKW, Trilegal and Cure, were also hired subsequently to draw up the specific terms of the contracts for private players and do a social and environmental assessment of the water supply in the city (Parivartan 2004).

The consultants' reports were released in April 2004, closely followed by the

DJB's *Delhi Water Supply and Sewerage Sector Reform Project* (DJB 2004), which laid out the road map for restructuring water supply in the city.

What was being proposed?

The *Delhi Water Supply and Sewerage Sector Reform Project* proposed:

- Management of each of the 21 zones of the DJB to be handed over to a private companies.
- The companies will not invest any money. They will simply manage a zone. This includes distribution of water in that zone, billing collection, grievance redressal, maintenance, etc.
- All DJB employees in that zone would report to the managing company.
- Water companies will receive a fixed "management fee."
- Companies would be given annual targets to achieve.
- Penalties to be imposed if companies fail in targets; bonuses if it succeeds.
- Water companies to supervise implementation of all capital works (Parivartan 2005).

Each distribution company was contract bound to ensure uninterrupted water supply at a specified pressure to the operating zone under its control. A fixed fee, called a "management fee,"[17] was to be paid to each company for running that zone. If it failed to meet the target specified for that zone, the company would have to compensate the DJB and its consumers in the form of penalties and lost bonuses (Parivartan 2004; 2005).

All payments for capital investments to be made by the private operator for infrastructure maintenance were also to be provided by the DJB to MNC. At the beginning of every year the company would provide an estimate of money required for capital works which the company proposed to undertake during the year. Theoretically, the DJB could vet this estimate, but in practice the DJB was obliged to make the amount available to the company or the company would be free of its obligations. No cap was prescribed on the amount that the company could demand. The company would be responsible for bidding, supervising, implementing, and certifying the completion and quality of work. The DJB had no control or say in the matter (Parivartan 2004). The company was not supposed to make any investments and the DJB's role was limited to providing the money the company demanded for infrastructure maintenance (Parivartan 2004).

While the DJB planned the implementation of the Delhi Water Supply and Sewerage Sector Reform Project (DWSSSRP) it had to simultaneously ensure that during the implementation period (2005–10), the following objectives set out in a World Bank Memorandum of Understanding had to be met:

Reliability. The DJB would (1) initiate "outsourcing" of the provisions of WSS services operation and maintenance (operation and maintenance (O&M) and commercial activities) to professional operators with the main objective of gradually moving from an intermittent supply system to a continuous (24/7) supply system in two operational zones (OZ), South I and South II; (2) systematically rehabilitate existing WSS infrastructure in these two zones; (3) selectively rehabilitate existing trunk WSS infrastructure in other parts of the DJB service area to reduce energy consumption and remove major bottlenecks; (4) implement a series of measures to improve the DJB's overall management performance; and (5) prepare a "roll out" plan for improving the WSS service in the entire DJB service area.

Sustainability. The DJB would: (1) gradually raise user charges so that revenues exceed cash O&M expenses by the end of year five; (2) clean its balance sheet and restructure its capital; and (3) reach full cost recovery in the South I and South II zones through a combination of efficiency gains and increased revenues. The Government of Delhi would gradually reduce operating subsidies paid to the DJB to fully phase them out by year five.

Affordability. The DJB would: (1) reduce its energy costs, currently constituting 42 percent of its O&M costs, by refurbishing its largest pumping plants; (2) reduce its establishment costs, currently forming about 45 percent of its O&M costs, notably by freezing recruitment and outsourcing selected tasks; (3) design a tariff structure that met simplicity and equity criteria in addition to financial objectives; and (4) implement specific WSS subprojects that would cater to the needs of low income communities (RFSTE 2005: 47–8).

Based on the World Bank's and the consultant's (PwC) recommendations, the Government of Delhi announced its plans to implement the project. While these plans were in the pipeline, the PwC recommended that, since total privatization in terms of handing over ownership was not acceptable, the same goal could be achieved through unbundling – giving management contracts for different activities to private parties – and through the Government raising tariffs as much as possible before the privatization commenced so as to protect the private players from public resentment (Raghu 2005).

Citizens Front for Water Democracy protest tariff hike

Based on the recommendations of PwC, the Government of Delhi announced a seven to tenfold increase in water tariffs on 30 November 2004 (RFSTE 2006). To discuss this immediate threat and to plan future actions, Navdanya/the Research Foundation for Science Technology and Ecology (RFSTE) called a meeting of the Citizens Front for Water Democracy (CFWD) in Delhi on 1 December 2004 at its office in New Delhi (RFSTE 2005). To oppose the unjustified hike in water tariffs, the CFWD, which is an alliance of women's organizations, environmental organizations, resident welfare associations, water workers, religious groups, student unions, farmer organizations, consumer groups, and health organizations, joined hands to oppose the DJB move to increase water tariffs. It demanded a revision of tariffs, fulfilment of social objectives and the defending of people's right to water (RFSTE 2006). It also demanded release of the white paper on which this price increase was based. It was decided to: (1) launch civil disobedience against this "cruel and unjustified" hike in water prices and call on all citizens not to pay bills at the higher rate; (2) undertake a citywide citizen's awareness program against the false basis of the price increase; (3) initiate a citizen's referendum on water tariffs (RFSTE 2006:25). To achieve the objective of reversing the tariff hike, the CFWD decided to:

1 Broaden the alliance.
2 Mobilize schools and colleges and religious leaders to oppose the hike.
3 Ask NGOs, resident water associations, disabled groups, and women's organizations to write letters to the Chief Minister about the water tariff hike.
4 Prepare posters/pamphlets against the hike.
5 Distribute badges with the message "Water is life, not for profit" (RFSTE 2006: 25).

While the opponents protested the tariff hike, some of the world's biggest MNCs had already entered Delhi, many of which are on the Fortune 500 list. Four MNCs: Manila Water (led by Betchel), Degremont (a subsidiary of Suez), Veolia (also called Vivendi) and Suar were short listed for water distribution in the South I and South II zones in early 2005 (Parivartan 2005). It was expected that with the short listing of these companies the project could be implemented by November 2005. However, media reports led to widespread opposition of the Delhi water reform project from a broad range of civil society groups, including trade unions like the Water Workers Alliance (WWA), community organizations of resident welfare

Table 5.5 Chronology of major events in the Delhi Water Supply and Sewerage Sector Reform Project (DWSSSRP)

Date	Events
1998	The World Bank enters Delhi and the restructuring of public utilities. Creation of the Delhi Jal Board (DJB) coincides with the Government of Delhi's clearance for the Sonia Vihar project to augment and treat water for Delhi.
2000	The DWSSSRP receives a preparation facility advance of $2.5 million from the World Bank.
2001	B World Bank consultants PricewaterhouseCoopers (PwC) draw up proposals for privatization, including contracts for the Sonia Vihar and 24/7projects, water tariffs and water legislation.
21 June 2002	Sonia Vihar Plant inaugurated by the Delhi Chief Minister and the contract awarded to Ondeo Degremont. Terms of contract kept secret.
8 August 2002	Ganga Yatra – a people's movement against the diversion of Ganga water to Delhi – declares "Our Mother Ganga is not for sale." The Water Liberation Campaign (WLC) alleges privatization of water.
2003	The Sonia Vihar Plant cannot be commissioned due to Governmental delays and opposition from people's movements.
22 March 2003	Citizens Front for Water Democracy (CFWD) to stop Delhi's water privatization launched.
15–22 March 2004	Jal Swaraj Yatra (water democracy journey) organized by the WLC against water privatization in Delhi.
Mid-2004	The DJB tries to run the 24/7 project.
Late-2004	Timetable prepared for implementation. The DWSSSRP report is released. The DJB is to begin implementation in 2005 and complete it by 2015.
November 2004	Ganga Yatra (Ganges journeys) and protests against water privatization take place.
30 November 2004	The Government of Delhi announces a seven to tenfold tariff hike. The CFWD protests the hike.
Early 2005	The DJB shortlists companies for the 24/7 project. NGOs and the Water Workers Alliance (WWA) criticize this as a move towards total privatization. People's movements rally against it. Letters written to the Prime Minister and Sonia Gandhi by intellectuals and NGOs.
September 2005	The Government of Delhi writes to the World Bank stating they have put the 24/7 project on hold.
23 November 2005	The Government of Delhi writes to the Government of India that they are withdrawing their World Bank loan application.
August 2006	The Sonia Vihar Plant, run by Ondeo Degremont, starts functioning as a Build-Operate-Transfer (BOT) project.
Current status as per reports	While the World Bank loan was withdrawn, the DJB authorities have stated that they will implement many of the recommendations by PwC and the 24/7 project. They are ambiguous on whether or not some "outsourcing" would be entailed in the projects.

Source: Compiled from newspaper reports and publications of the Research Foundation for Science Technology and Ecology (RFSTE).

associations, environmentalists, consumer organizations, and political parties. Since water is under the control of the individual Indian states, in most cases the protests were localized and diffused, without parties forming a single national alliance. But since Delhi hosts the capital city, the reform project attracted more attention and brought about protests from all over India. The media, too, gave it national publicity. Most of these protests and campaigns were initiated by local or national organizations under the banners of the WLC and CFWD. The WLC consisted of NGOs[18] like Paani Morcha, headed by Commander Sureshwar Sinha; Tarun Bharat Sangh, led by the "Water Man," Rajendra Singh; Dehat Morcha; the Bhartiya Kisan Union (a farmers' organization); and Bhartiya Jagriti Mission (a religious organization).

The Right to Water Campaign, the CFWD, the Delhi Water Sewer and Sewage Disposal Employees Union at the DJB, the WWA, and others conducted relentless public education and protest campaigns. The aim was to educate fellow residents and to put pressure on the Government of Delhi, the Government of India, and the World Bank country representatives. Cross-class alliances were forged between the middle class residents of posh residential colonies and poor residents of slum clusters (Sehgal 2007). A common set of concerns and demands was constructed after several rounds of meetings to ensure that everyone could come together on a common platform to oppose the reform plans under the guise of providing water 24/7 to Delhi's residents (Sehgal 2007). A campaign that began with a handful of people grew to hundreds of participants debating in community halls and finally spilled onto Delhi's streets with thousands of protestors joining the campaign from all over the country.

Widespread public protests were made over the much publicized PPPs that ranged from MNCs taking over Delhi's water for profit, the role of the World Bank in facilitating private sector participation in water, and tariff hikes as a step toward accomplishing this process of what many called "back door privatization." Campaigns revolved around a broad set of interests based on social, economic, and environmental grounds. The local communities, the urban poor and women, who would lose access to water if the proposed reforms were implemented, were very active in the campaigns. Other issues included challenging Government claims of reliability, efficiency, public involvement, and transparency. The protests also drew attention to the effects privatization would have on the common man and woman. The demonstrations were part of the agitation against water sector reforms undertaken by the Government of Delhi following the guidelines of the World Bank. Activists argued that the city was witnessing a major thrust toward privatization in the name of "reforms."

The government and proponents of the free market, however, advocated the water problems of Delhi as issues of wrong pricing, inadequate incentives, free riders, subsidies, and inefficiency. The solution according to them was to enable the rule of the self-regulating market in urban water supplies and services.

This brief history of the water sector in Delhi conclusively shows that shifts in the urban water policy provisions have occurred. Before the 1990s, water fell under the provisions of public sector undertakings but processes of liberalization brought about a paradigm shift in the perception of urban water supply in the states of India.

In the case of Delhi, the Government of Delhi announced its reform program based on consultants' reports that recommended the management of water with a rational and technical approach. Protestors on the other hand claimed that these decisions masked the real intent of privatizing water supply. These claims and counterclaims that emerged out of the reform process are discussed in the following chapters. Chapter 6 and 7 explore the tactics that these different actors adopted in an attempt to shape the direction of policy through the shifting narratives of the proponents and opponents of policy reform that both produced and drove the policy process.

Notes

1 Fueled by the economic boom of the 1990s, India's capital city has emerged as the country's favored place for opportunity seekers. It is today the largest destination for foreign investment, the biggest market for consumer products, the most active center for art and culture, offering a radically changed life-after-work.
2 A city with first world desire is one based on the Western model of development that would include super highways, malls and the infrastructure and material aspirations of the West.
3 The road network increased from 8,380 kilometers in 1971–72 to 31,183 kilometers in 2005–6. A special program to construct flyovers was started in 1998–99 when Sheila Dixit took power and 34 flyovers had been completed up to 2005–6. The first phase of the Metro system was completed in 2006 and second phase is expected to be completed by 2010 (see http://www.india inbusiness.nic.in/no-india/states/delhi.htm (accessed on 30 August 2007)).
4 In this respect, it is like Chicago which is located at the connection point between the Missouri-Mississippi River System and the Great Lake System (Watkins 2001).
5 By optimization, about 675 MGD water is produced (DJB 2004: 1).
6 Internationally, the term "water loss," better known as "nonrevenue water," represents water that has been produced and is "lost" before it reaches the customer (either through leaks, through theft, or through legal usage for which no payment is made). Part of this "lost" water can be retrieved by appropriate technical and managerial actions. It can then be used to meet currently unsatisfied demand (and hence increase revenues to the utility), or to defer future capital expenditures to provide additional supply (and hence reduce costs to the utility). (An international benchmark network for water and sanitation utilities toolkit is available online at: http://www.ib-net.org (accessed 12 July 2007)).
7 One crore is equal to 10 million rupees and 100 crores is equal to 1 billion.

8 This process in World Bank language is known as "unbundling," the goal of which is to reduce monopolies as much as possible and subject everything to market forces. This neoclassical idea is endorsed wholeheartedly by the World Bank in infrastructure in general, and water in particular (Finger and Allouche 2002: 73–4).

9 This plant is being run by the multinational Suez under a ten year Build-Operate-Transfer (BOT) contract.

10 Under the terms of the contract all performance targets assigned to contractors were based on a presumption of adequate raw water supply by the DJB. If the supply proved inadequate, which is most likely in the given scenario, the contractor was absolved of all performance obligations and targets (Sethi 2005). The Chief Minister of Delhi, Sheila Dixit, reported in an interview that the Sonia Vihar Plant had not been able to solve the water problem as it has not been receiving adequate supply of raw water from UP.

11 The construction of the plant was cleared by the Government of India in 1998, specifically by the Central Public Health and Environmental Engineering Organization.

12 The terms of the contract were kept secret by the DJB. Even after several petitions by NGOs under the Right to Information Act the DJB had not posted the terms of the contract on its website by 1 September 2007. Information available on the terms of the contract was leaked to the press, and protests against the company began from there with the first major rally on 8 August 2002 organized by the Water Liberation Campaign adopting the slogan "Ganga is not for sale."

13 This argument was raised in a report written after a group of experts visited a similar Degremont plant in Surat on the river Tapi and expressed doubts about its success (personal interview with Suresh Babu, Centre for Science and Environment, New Delhi).

14 The DJB is already laboring under a huge debt of 33 crores due to the delay of the commissioning of the plant in November 2006.

15 A full chronology of the major events in the reform project is given in Table 5.5.

16 The DJB approached the World Bank in 1998 for a loan. The World Bank's team visited the Board in July 1998 and recommended they hire a consultant who would "suggest" basic reforms for the DJB to carry out. The World Bank offered a $2.5 million loan to DJB to hire a consultant. The loan carried an interest rate of twice the amount the Government would have had to pay if it raised the money from the internal market. Besides, the loan formed just 10 percent of the amount that the Government of Delhi spends on DJB every year (Parivartan 2004). The process was finalized in complete secrecy. Reports appeared in the media of the World Bank's role and involvement in awarding PricewaterhouseCoopers (PwC) a Rs 7 crore ($ 1.6 million) consultancy contract in November 2001. (For details of the controversy see the Parivartan report, *Delhi Water Supply and Sewerage Project: An Analysis* (2005).)

17 Three types of payments were to be disbursed to the private company: (1) management fees to meet the salaries of the employees sent by companies; (2) operational and day-to-day expenses to run the zones where the companies are functioning under the DJB contract; and (3) capital investments to make improvements and maintain infrastructure. Credit goes to the NGO Parivartan for exposing the details and underlying implications of this project.

18 Paani Morcha is an organization that works to alleviate the growing water crisis in India particularly in Delhi through public interest litigation and for the liberation of Ganges water. Tarun Bharat Sangh is an NGO that brings people together on issues of management of forests and water resources. Dehat Morcha and Bharatya Kisan Union is a farmers' movement which fights for the rights of farmers over land and water. Bhartiya Jagriti Mission is a religious and charitable trust working for the welfare of rural poor and to save the water of Goddess (as their website calls her (http://www.gangamaa.com)) Ganges. Their main goal is to confront the challenges of pollution, privatization and construction of the Tehri Dam on the Ganges. The Citizens Front for Water Democracy was created by these organizations and the people of Delhi to fight water privatization. Resident Welfare Associations were formed in every housing locality to

take care of the welfare and basic needs of the residents in that locality. Water Workers Alliance is a group formed by the middle and lower rungs of bureaucracy in the DJB to fight water privatization. The National Federation of Indian Women is the women's wing of the Communist Party of India to intervene and interact with policy makers to help women get their rightful place and rights in society.

6 Mainstreaming policy
Discourses of power

An overview of the major shifts in the discursive construction of water, as well as the policy shifts, which have been associated with different stages in the evolution of the discourse, was presented in Chapter 4. The way the concept of water is framed has an important influence on the ways in which water reform policies come to be shaped. Shifting narratives of the causes of and solutions to water issues both produce and drive policy processes, making spaces available (as well as reducing them) in which different forms of water knowledge can be articulated and mobilized. These narratives not only convey storylines of causes and consequences but also often have embedded within them the advocacy of particular policy instruments: "What appears as knowledge is structured by the aim to which it is directed" (Nustad and Sending 2000: 221). Constructed in a way that permits intervention, the promotion of particular technical approaches lends further persuasiveness to policy discourses. When storylines are appealing and instrumentation appears to be clear-cut, uncertainties are dispatched and all that is required is implementation.

By focusing on how water is positioned among the mainstream actors, this chapter sheds light on how knowledge in the policy processes for water reform became constituted as useful and legitimate. It reveals those constructions of water that captured the attention of mainstream actors – the World Bank, PricewaterhouseCoopers (PwC), the Government of Delhi and the Delhi Jal Board (DJB) – citing policy documents and interviews as evidence of the way these actors framed their concepts of water.

The World Bank

The World Bank's website (http://www.worldbank.org) reveals certain key facts about its role in the Delhi water reform project. The website claims that Delhi has an adequate supply of water. With over 200 liters of water available per capita per day, Delhi has more water than many other big cities in the world that provide their residents with a 24/7 water supply. Despite this apparent abundance, the website asserts that in its current mode of operation, the DJB is unable to meet the water and wastewater needs of the nation's capital and provides its citizens with an erratic and unequally distributed water supply that is well below international standards. The World Bank's website claims that the Government of India and the Government of the National Capital Territory (NCT) of Delhi recognize the urgent need for reform and have requested the World Bank's support in helping the DJB improve the reliability, sustainability, and affordability of Delhi's water supply and sanitation services.

In 2005 Michael Carter, the World Bank's Country Director for India, stated that the World Bank's goal in the Delhi water supply project was to ensure a continuous and regular supply of water to the citizens of Delhi, including the urban poor in the Jhuggi Jhopri (JJ) clusters (Carter 2005). The World Bank, whose main goal is the elimination of poverty, feels that water reform will set right the problems of scarcity, quality, and regularity in water supply to Delhi. It proposes to achieve this through a public–private partnership (PPP), not, it argues, for the purpose of privatization, but for efficiency and proper management of water through infrastructural development and quality assurance. Carter categorically asserted that "Neither under the proposed project nor in any advisory work is the Bank proposing privatization of any part of DJB nor is there a timetable for privatization. As a matter of fact, at this time, the World Bank would definitely not recommend privatization" (Carter 2005).

Elaborating on some of the important dimensions of the World Bank's role, Carter stated that the management of water distribution was to be accomplished by

> delegating the operations and maintenance of two operational zones to experienced private operators under two separate management contracts with DJB. The operators will be selected through a competitive process according to the Bank's procurement guidelines. DJB will retain control over assets, staffing, tariff and investment decisions, and will supervise the operator. The operator will receive a fixed fee with bonuses and penalties depending on performance.
>
> (Carter 2005)

The project's goal, the World Bank proclaimed was to achieve a set of standards for better delivery services in the water sector and efficiency and economy in financial management which required changes in the current water management processes:

> This requires setting tariffs at a level that covers at least the cost of operations and maintenance. This is a matter we need to reach an agreement with DJB on. As an integral part of its support to DJB, the Bank will continue to ensure that all procurement that it finances is done efficiently and in a transparent manner.
>
> (Carter 2005)

In the FAQs section of its website in 2005, the World Bank emphasized the need for involving foreign companies in Delhi's water reform:

> Upgrading a system to provide continuous water supply requires specific expertise. As developing countries across the world face similar problems, many companies have successfully tackled these challenges and have acquired the necessary experience. Bringing in operating expertise and know-how can therefore upgrade DJB's operations and train the utility's staff in management methods that are on par with world standards.

On the accountability of these companies, the World Bank claimed in the same section of website that:

> Clear performance criteria will be established and recorded in enforceable contracts. DJB will pay the operators a fixed management fee for meeting the specified performance criteria. The management fee will be determined through an open, competitive process. Operators will be paid a bonus if they exceed the minimum criteria, and penalties will be imposed for poor performance. Operators will be contractually required to share part of their bonus with DJB employees seconded to their operations.

Again the FAQs section, the World Bank also asserted its main goal as elimination of poverty:

> The proposed reform specifically covers the poor. JJ clusters in the two zones will have household connections through shared meters. The operators will also be obliged to set up a Poverty Outreach Unit in both zones and will be contractually required to improve water supply in poor settlements. Continuous water supply will not only ease the burden on households and

provide them with safer water, but will also reduce their costs on buying water from tankers, making extensive arrangements for water pumping and storage, and buying their drinking water or purifying it themselves .... Today, while the poor may be the intended beneficiaries of the low charges, they suffer the most from the poor quality of service that results. As the quality of service improves, customers should have to rely less on expensive "substitutes" such as boosters, storage tanks, and purification equipment.

Sumir Lal (2005), the World Bank's Advisor, External Affairs, has described his own experience and the plight of citizens of Delhi who get their water supply of about 250 liters per person per day for only a few hours a day:

> Ask the poor person who must line up for hours for a water tanker, pay five to eight times the official tariff to a private supplier, and then watch his or her children repeatedly fall prey to diarrhea and hepatitis. Or a widow that gets up at 4 a.m. to fill her bucket. Suppose instead that DJB plugged its leaks and found an efficient way to deliver continuous and safe water supply to all its citizens, including the poor who are excluded now and charged a reasonable tariff to cover its operating costs; while the government subsidized poor consumers and picked up the tab for DJB's capital investments. Surely, a preferable scenario?

This is the model, he argued, that the proposed DJB reform project, which has run into so much controversy, seeks to provide:

> No privatization. No Fantastic tariff hikes. No "World Bank model." Simply, a system that works in dozens of other cities in developing countries, even in those with terrible governance environments. Why must India alone persist in denying its citizens access to the most fundamental life good – clean water?
>
> (ibid. 2005)

Lal argued that given the wealth of engineering talent in India, the saner assumption would be that these outside companies would deploy a mix of expatriates (to bring in the know-how, and the technical expertise) and resident Indians; and if the experiment does get scaled up, it would largely be resident Indians, not expatriates or foreigners who would be hired. In the process, an Indian water management industry, which is currently absent, would develop.

The World Bank and its representatives couched their discourse in terms of providing safe and potable water to all, especially the poor, through PPPs that ensure reliability, sustainability, and affordability based on the knowledge and expertise that their experts possess in the area of water management. The discursive package through which private sector participation is constructed is called "water resources management." The rationale behind this approach holds that the private sector can be used effectively to manage waters to enable sustainable development and alleviate poverty. According to the World Bank, breakdowns in the system have resulted from bad management practices by the state in maintaining infrastructure networks and, given the World Bank's belief in the efficacy of market forces, management reform from its point of view should involve a wider application of commercial principles to service providers. From the "water resources management" point of view then, a program of cost recovery and greater use of pricing can be expected to lead to improved performance and sound financial and economic management (Carter 2005).

The World Bank's approach to water speaks to the efficient management of water resources as a key to economic growth and development. Its discourse does not focus solely on poverty alleviation and economic growth. In fact, it selectively incorporates concepts generated by alternative discourses and some assumptions, for example, using terms such as "public consultations," "partici-patory development" and "people-centered."[1] The use of such carefully chosen key words in their narrative, however, serves only to qualify, not to redirect, its fundamentally economic perspective. Water is still framed as an economic problem whose solution must therefore be economic as well. The need for capi-tal investment and economic growth to bring the needed improvement is thus presented as a self-evident truth. The World Bank presents its proposed solution to Delhi's water supply problems in technical and politically neutral terms. Its policies prioritize the construction of physical infrastructure, technological qual-ity, and good governance, which are seen as the first steps toward growth and sustainable development.

Based on the World Bank and PwC's recommendation, the Government of Delhi announced its plans to implement the project to manage water rationally and technically. Four multinational corporations (MNCs): Manila Water (led by Betchel), Degremont (a subsidiary of Suez), Veolia (also called Vivendi) and Suar were short listed for water distribution in South Delhi I and II zones in early 2005 (Parivartan 2005). It was expected that with the short-listing of these companies the project could be implemented by November of 2005.

The Government of Delhi and proponents of the free market however, presented the water problems of Delhi as issues of wrong pricing, inadequate incentives, free riders, subsidies, and inefficiency. The solution according to them was to rope in the private sector in the name of PPPs. The following section sheds light on how knowledge in the policy processes for water reform became constituted as useful and legitimate. It reveals those constructions of water that have captured the attention of the Government of Delhi and the DJB – citing policy documents and interviews as evidence of the way in which the Government frames its concepts of water.

Government of Delhi claims

The Delhi Water Supply and Sewerage Sector Reform Project (DWSSSRP) was intended to solve the problems of uncertainty in the bulk supply of water, intermittent and inadequate water supply, nonrevenue water, absence and ineffective metering and inefficient operations (DJB 2004: 8–12). The overall position of the Government of Delhi on reform was that it supported it. However, on closer analysis, we can uncover more nuanced, and at times conflicting interpretations. It is often in the voices of individual legislators and lower level bureaucrats that we find alternative arguments for reform. This section explores some of the dominant emerging themes from the Government as articulated by the DJB. It also explores the more nuanced visions of a minority of individual legislators and the Water Workers Alliance (WWA) who comprise those in the bureaucracy of the DJB that opposed this reform. The Government of Delhi, the central Government of India, the Planning Commission, the World Bank and the consultants all seem to share a similar position on water reform in Delhi. This situation strongly reflects Hajer's "discourse coalitions" where previously independent practices on infrastructure development in water are being brought together to give meaning within this common political project of urban water reform. However, the Government of Delhi also realizes that the success of the reforms depends not only on the discourse coalition but to a great degree on stakeholder participation.

The Government of Delhi's water reform project report, therefore, mentions the involvement of all the stakeholders, claiming that the success of the reform would depend on addressing their concerns. The report states:

> a workshop was held in March 2004 to formulate a Sector Vision in water and to set an agenda for the reform process. The workshop brought together representatives from DJB's management and staff, central and state government, resident welfare associations, multilateral and bilateral development agencies, NGOs, and experts from progressive water utilities on a common platform to deliberate on the issues facing the sector, share best practices, and reach a consensus on the proposed vision. To take the visionary agenda

forward, another workshop was held in May 2004 to develop a Reform Implementation Strategy. The objective was to bring together various stakeholders to deliberate on the strategy and implementation mechanism and develop a concrete action plan for undertaking reforms in the next five years.

(DJB 2004: 8)

Here we see the Government of Delhi and the DJB setting out to position themselves as communicators who make decisions based on consultative processes and discussions, and participation of stakeholders. The adoption of this position has strategic significance as it promotes a view of the decision making process as being based on consensus and hence uncontroversial.

By citing previous examples of PPPs in different developmental projects, the Government focuses on convincing the people that this reform project is not different from the others undertaken. The Chief Minister of Delhi, Sheila Dixit, justified private sector participation in the project of the DJB in terms of savings in cost and time; such participation would also ensure outstanding execution and professionalism:

Various road and highway projects are being executed on a BOT [Build-Operate-Transfer] basis, which is not new to the state. The people of Delhi should have no fear on this front. We are not going to privatize drinking water supply. We are going in for use of modern technology which would make life much easier for the consumers and help in regulating the supply of water to various parts of the city.

(*The Hindu* 2 October 2004)

The semantic shift that she uses in justifying the proposal is the term "outsourcing" instead of "privatizing": "The Planning Commission has asked us to make the entire procedure for outsourcing the management of water supply more transparent" (*Express News Line* 2005:1).

The Government of Delhi also strongly emphasized that these private players will be fully accountable to the Government. In consonance with the World Bank objectives, the Government mentioned the establishment of clear performance criteria to be recorded in enforceable contracts. Using the same language, the Government claimed that a fixed management fee will be paid by the DJB to the operators for meeting the specified performance criteria. Operators would be paid a bonus if they exceed the minimum criteria and penalties will be imposed for poor performance. The Government therefore claimed that its proposals are about management and distribution of services through PPPs[2] and presented these measures as steps toward a sustainable, safe and continuous water supply for its people.

Official categorization of water supply and services

The Government of Delhi justified its proposal based on the official categorization of water supplies and services in Delhi in its report (DJB 2004). The report asserted that the actual water supply available to the residents "is intermittent and inequitable"; the demand–supply gap is on the rise; losses are around 40–50 percent; and intermittent supply leads to increased health risks from possible contamination of leaking pipes. Furthermore, there were shortcomings at treatment works and the equipment is inefficient. The obvious consequences of the above situation, the report claims is, "poor reliability, increased health risks due to inadequate water supply and management of wastewater, huge coping costs and low customer satisfaction" (DJB 2004: 2, 7–8). These quotes are an example of how such categorization was used to demonstrate that priority must be given to public–private management practices in the decision making process.

The Government of Delhi's position and national and international policy frameworks

The DJB report argues that the proposed reform project is in line with the Government of India's National Water Policy (NWP) (2002), which advocates an "integrated and multidisciplinary approach toward water resource planning, development, conservation, and management" (DJB 2004: 10). The NWP (2002) lays emphasis on the formulation of state water policies backed with an operational action plan in a time bound manner within two years. With respect to the policy aspects:

> The NWP lays emphasis on the need for the paradigm shift from development to effective management of water resources through recognition of water as an economic good with well targeted and transparent subsidies for the poor, adoption of scientific water management techniques, participatory approach to water sector management and encouraging private sector participation.
>
> (DJB 2004: 10)

The report claims that these principles were incorporated in the proposed reform project. The DJB also justified its position on the grounds that, "the project is in line with the 74th Constitutional Amendment, which aims at devolution of enhanced financial and functional powers to the urban local bodies to enable them to function as effective self-governing institutions. It mentions that the project is consistent with India's Tenth Five-Year Plan, which recognizes that infrastructure bottlenecks have become a major constraint on growth, and therefore poverty alleviation. The report claims that access to safe water has been declared as a basic human right by the United Nations and is widely recognized to contribute directly to poverty alleviation. This approach, the report asserts, would fulfill those goals and would also help to achieve the Millennium Development Goals adopted at the World Summit on Sustainable Development in Johannesburg in 2002, which laid emphasis on the provision of safe drinking water, improved sanitation services, and sustainable environment management" (DJB 2004: 10–11).

The DJB and the Government of Delhi rationalized these reforms on the basis of various policy pronouncements of the Government of India, such as the NWP (2002), the 74th Constitutional Amendment, and the Tenth Five-Year Plan (see Chapter 4), arguing that the policy change was in line with national goals. The Government of Delhi also justified the policy change through the Millennium Development Goals (MDGs) adopted by international institutions like the United Nations and the World Summit on Sustainable Development, which hold that access to safe water and improved services contribute to sustainable development and the alleviation of poverty. The Government of Delhi thus demonstrated that it was fulfilling a national and international vision for continuous water supply, better quality and improved services to the poor, environmental and financial sustainability, operational efficiency and accountability.

Technology, quality, efficiency and maintenance as the reform project's outcome

The Government of Delhi's position was in keeping with the vision of Delhi's water policy in which private companies would be recruited to ensure both bulk supply of treated water and a regularized supply of distribution management. The chief executive officer (CEO) of the DJB, Rakesh Mohan, claimed that buying private expertise in management is no different from buying pipes, pumps and meters (*The Economist* 2005). Considering the fact that the DJB made a conscious decision to upgrade services, increase responsiveness, incentivize demand management and increase awareness among people about the value of water in order to promote conservation, the Government of Delhi argued that it needs to professionalize its service and adopt best international practices in operations and management of the water sector.

In a personal interview conducted on 6 December 2004, a DJB official explained that the Sonia Vihar project (see Chapter 5) was in line with the Government of Delhi's vision of potable and continuous supply. He said that the contract was awarded to Ondeo Degremont for reasons of technology, quality, and efficiency:

> We do not possess this technology in India. Operation is inefficient – and maintenance is at far lower levels than is needed; we want to improve our delivery system. If I feel that I am not confident at the managerial level or delivery level and I want to improve it, I take advice, managerial help or consultancy. How can you call it privatization?
>
> (Personal Interview 6 December 2004)

In the same interview, the bureaucrat compared the Sonia Vihar Plant to a cricket or football team, pointing out that many teams in India have foreign coaches:

> We hire a foreign coach for efficiency and expertise that he can give to the players. If we have a foreign coach it does not mean that we have privatized our team. The same goes for Ondeo. Ondeo provides us the expertise of technology and efficiency of operation but that does not mean that the plant is in private hands. It is just improving the system for efficient, equitable, and sustainable management.
>
> (Ibid.)

Similar arguments are made in the DJB report. The technical and managerial expertise of the private sector would be used to achieve significant efficiency gains and to train the DJB employees in modern utility management best practices (DJB 2004: 3).

The Government of Delhi claimed that the pilot project run from the Sonia Vihar Plant in the Delhi zones South I and South II would provide 24/7 water supply by the use of modern technology that would come from these private companies. The main purpose of utilizing these companies was to regularize supply and plug the loopholes through their technological expertise. Private firms would run the system, but the chief minister of Delhi refused to term it privatization of water distribution. "We are just taking the help of some experts in the field of water distribution as the DJB does not have expertise to operate such a system," she said *(Hindustan Times* 30 September 2004). Thus, the claims of both politicians and bureaucrats are based on technology and expertise, highlighting the shaping of water policy reform as a Foucauldian "political technology" and relying on the versions of "expertise" and institutional techniques that create and define the category "water."

Cost recovery essential for better services and financial sustainability

The DJB report addressed the problems of low tariffs, low cost recovery,[3] and consequently, the reliance on excessive loan assistance from the Government of India, the high cost of water supply, and energy usage. Even though Delhi has the highest national per capita income, it had the lowest water tariffs among all the cities in India. While the 1998 Act that constituted the mandate of full cost recovery, this philosophy has not been reflected in the tariff-setting decisions taken by the DJB and water prices have remained low and subsidized. According to the Government of Delhi this resulted in a persistent deficit on the revenue account and a recently conducted Willingness to Pay Survey in Delhi indicated a positive response to the idea of more charges for better service and quality. The Government justified the project based on numbers and statistical calculations of a people's willingness to pay:

> The Government claims that given DJB's envisaged pipeline of projects, funds required for system rehabilitation, as well as limited capacity of the Government as the sole funding source, there is a need to progressively revise water charges accompanied with improved services whilst gradually phasing out the Government subsidy, with the objective of achieving full recovery of costs of efficient operation and maintenance.
>
> (DJB 2004: 9)

Water tariffs were hiked in a process that came into effect on 30 November 2004. The CEO of the DJB, Mohan (RFSTE: 2005), explained that the pre-revised water tariff in Delhi was amongst the lowest in the country. This water tariff increase, he argued, was intended to ensure that there is no water wastage and to increase the participation of consumers in following practices for saving potable water:

> With the population of Delhi growing everyday and with no other addi-tional water sources in sight, it is important that the consumers use water for priority uses and tariff is only one of the tools being used for demand

management. Raw water to the treatment plant ... was being brought at huge costs, and costs were also being incurred to treat the water to potable quality and thereafter in maintaining its quality and ultimately in collection, conveyance and treatment of that water.

(RFSTE 2005: 2–3)

Recovering the total operational cost from the consumers was not the goal, "but the consumers are being asked to share in the cost of maintenance of the systems, which are being laid to take the potable water to their doorstep" (ibid.: 3). The Government targeted managing water demand through cost recovery programs of higher tariffs in the name of providing better and reliable services.

Reform project to benefit the people of Delhi and provide targeted benefits to the poor

Arguing "while the poor may be the intended beneficiaries of the low charges," the DJB report claimed that they suffer the most from the poor quality of service that results. As the quality of service improves, customers should have to rely less on expensive "substitutes" such as boosters, storage tanks, and purification equipment" (according to the FAQs page of the World Bank website in 2005). The Government of Delhi justified its proposal on the grounds that all institutional and technical interventions are designed with a clear obligation to improve the service to the poor. Challenging the "myth" promoted by the opponents of water privatization that "eventually the poor will be deprived of their rights of water," the DWSSSRP aimed at ensuring access to the economically weaker sections of the society through specific pro-poor measures.

According to Government of Delhi officials, the DWSSSRP aimed to provide piped water to the poor in the resettlement colonies, slums, and JJ clusters where people are most susceptible to disease and have to fetch water from long distances. Improving water supplies and sanitation to the poor remains an urgent priority. Most poor reside in settlement colonies, JJs, and urban and rural villages that are partially covered by the water distribution system, and the DWSSSRP intended to meet their needs with targeted interventions:

The proposed reform specifically covers the poor. JJ clusters in the two zones will have household connections through shared meters. The operators will also be obliged to set up a Poverty Outreach Unit in both zones and will be contractually required to improve water supply in poor settlements.[4]

(DJB 2004: Introduction)

Pilot projects to be undertaken for improving water and sanitation services to the poor were to be selected based on such considerations as the technical feasibility, financial viability, legal aspects, ability to mobilize, and management capacity of the communities and then be institutionalized based on the lessons learned (DJB 2004: 17).

Ultimately, the Government of Delhi argued that continuous water supply would not only ease the burden on households and provide them with safer water, but would also reduce their costs of buying water from tankers, making extensive arrangements for water pumping and storage, buying their drinking water or purifying it themselves. According to the DJB report (DJB 2004: 23–4), addressing the water supply and sanitation sector would have a visible impact on multiple fronts such as poverty alleviation, environmental sustainability, public health, life expectancy, participatory development, and good governance. Even though the net benefits were difficult to quantify, they could be expected to be positive. The report envisaged that the DWSSSRP would be the driving force in Delhi to turn around the current situation of general poor performance into sustainable one where there were high quality services for all. The targeted beneficiaries of the various components of the reform measures would be the urban population of Delhi, totaling about 14 million. The Government claimed that tangible benefits would result in economic, social and environmental areas.

These stated claims of the Government of Delhi represented its case for PPP in Delhi's water sector. The conditions of uncertain and intermittent water supply, excessive nonrevenue water, and poor reliability due to inefficient equipment and other shortcomings at treatment plants are utilized to explain increased health risks from possible contamination and allusions are made to social issues such as inadequate services to the poor. These systemic inadequacies, the DJB argues, combine to produce high cost and low customer satisfaction.

The Government of Delhi essentially adopted the outlook of PwC. Its report stated that "despite having abundant water as well as sufficient treatment capacity, the actual service is poor" and requires "managing development through demand management." The Government developed a vision for the water supply and sewerage sector – "the provision of universal 24/7 safe water supply and sewerage services in an equitable, efficient and sustainable manner by a customer oriented and accountable service provider (DJB 2004: 9) – but if this was to occur, it argued, the service must be professionalized and best management practices, as defined by PwC, must be adopted in the form of private expertise and cost recovery for better services and the financial sustainability of the sector.

The Government of Delhi defended the World Bank's statements and reports, saying they would help the DJB in executing the project in a professional manner.

Emphasizing the desperate need to modernize Delhi's water distribution, the Chief Minister of Delhi, Sheila Dixit, remarked in a statement reported by the Indo-Asian News Service (22 August 2005):

> Delhi is an old city – leakages in water supply result in huge losses, leading to shortage. And distribution itself is uneven, ranging from a supply of 500 liters per person in some areas to less than 32 liters per person in others. To remove such anomalies we asked the World Bank to prepare a report – where does it mean that we are going to privatize water in the capital?

In following the lead of the World Bank and PwC and believing that such reforms can best be achieved through PPPs, the Government of Delhi demonstrated its involvement in a strong science–technology and industrial network that would provide the necessary expertise and capital required for such projects. With the use of the science and technology that private companies possess, the Government aimed to replace the intermittent and irregular supply of water facing the residents of Delhi today with universal 24/7 continuous supply. The Government's narratives of efficiency, quality, health, and environmental sustainability which it promoted in its report touch on major contemporary public policy issues both globally and nationally, and are thus reproduced in the neutral language of good governance.

Dissenting voices

While the Ministry of Water Resources, the Ministry of Urban Development, the Planning Commission, Delhi, the NCT Government of Delhi and the DJB presented a unified front on the issue of water reform, there were a few politicians and a section of the bureaucracy at the middle and lower levels that formed the Water Workers Alliance (WWA) with dissenting perspectives on the issue at both the local and state levels. The developmental vision of the Government was highly contested. Its vision of the future of development in Delhi is technocratic, based on foreign investment, BOT projects, and infrastructure development, of which water is a critical part. Opponents of this vision included not only the bureaucrats and political parties that favored a nationalist–protectionist discourse but intellectuals and activists that came from a broad range of civil society groups, including trade unions like the WWA, community organizations of Resident Welfare Associations, environmentalists and consumer organizations. Together they combined to form the Water Liberation Campaign (WLC) and the Citizens Front for Water Democracy (CFWD) alliances.

Widespread public protests were made over the much publicized PPPs that ranged from MNCs taking over Delhi's water for profit, the role of the World Bank in facilitating private sector participation in water, and tariff hikes as a step toward accomplishing this process of what many called "back door privatization." Campaigns revolved around a broad set of interests based on social, economic, and environmental grounds.

Newspaper reports carried claims by politicians from the Legislative Assembly that the Government of Delhi was betraying the interests of the common people for profits and running the Government as a corporate entity. Politician Madan Lal Khurana claimed that there was a deliberate attempt by the DJB to undervalue its assets and overvalue its services so as to allow MNCs to take over the assets and utilities at a very low price (*The Hindu* 21 April 2005). The Delhi Legislative Assembly members demanded an explanation from the Chief Minister and criticized the tenfold tariff hike in Delhi, which they claimed would hit middle and lower class consumers.

While politicians and members challenged the Government of Delhi's claims on costs, a section of the bureaucracy that consisted of engineers and the lower rung of officers in the DJB's hierarchy defected from the Government's position and in protest formed the WWA. This alliance, led by Naqvi, an engineer of the DJB, later joined with the Water Liberation Campaign to stall the process of privatization in Delhi. Being in the Government themselves, they said that the DJB's financial sustainability could be achieved through public–public partnerships and cooperative initiatives. Based on research entitled *Delhi Jal Board Financial Sustainability is Possible through Public–Public Partnership* which they submitted to the Government on 8 September 2004, they claimed that there was no need to burden citizens with enhanced loan liabilities as there was huge potential for the new raised tariff structure to meet the DJB's requirements (Sharma, *et al.* 2004).

They also presented a plan that could augment water supply and, in partnership with the NGO Navdanya/RFSTE and the WWA, claimed they could run the Sonia Vihar Plant without Ondeo Degremont. The proposal was to run the Sonia Vihar Plant by recycling wastewater discharged from the six water treatment plants operational in Delhi and by recharging the natural reservoirs in the Yamuna Basin during the forthcoming monsoon season.

According to Sanjay Sharma of the WWA:

We have already prepared a feasibility report and even worked out other technical details and scientific studies. The Government will not have to spend extra money and in turn, earn over Rs. 600 crores annually as revenue from sale of water.

This way, 195 MGD [million gallons per day of] water would be available which can be used for running the Sonia Vihar plant as well as to meet the requirements of treatment plants at Nangloi and Bawana. The immediate availability from the three treatment plants would be around 43 MGD that can be enhanced to its optimum level by August 2006. Transporting water to the plant would not be a problem as the river Yamuna would become a channel, saving the Government any extra expenditure on laying pipelines or setting up infrastructure.

(The Hindu 6 April 2005)

These groups sympathized with the activists and argued that the programs and the policies of the DJB were positive indicators towards the privatization of urban water supplies of Delhi. They spoke out against transnational corporate players and even provided alternative plans to the Government for running the system, augmenting water supply, and making the DJB financially sustainable (Sharma, *et al.* 2004). These tensions within the levels of government institutions provide evidence that the Government of Delhi is not a monolithic entity as understood in traditional policy studies but an area of conflicting aims and tensions within which interactions take place. Many on the middle and lower rungs of the bureaucracy expressed reservations about the Government's water policy, yet the dissident voices remain unheard and muted within the corridors of power while global thinking was consistently reflected at the national and subnational levels of policy making in India.

Analysis of these discourses emerging out of the Delhi water reform project show how an atmosphere of crisis was established through narratives of the irregularity, poor reliability, and scarcity that dominated the water sector in Delhi which necessitated water reforms in the first place. A network of experts – not only technicians but also administrators, strategists, and political advocates – was mobilized to drive this vision forward. The global expertise came through the "invited participation" of the consultants PwC, who were engaged to assess the status of the water sector and to provide inputs for remedial measures. These consultants were hired with the approval of the World Bank which was funding the project. Economic growth facilitated by technical experts and national and state Governments engaged in a donor–recipient relationship would result in good governance and good infrastructure leading to poverty alleviation and growth.

The Government of Delhi echoed the rhetoric of the World Bank and the consultants, citing efficiency, quality of water, and 24/7 water supplies to the poor via the science, technology, and skill of private players. This thinking echoes the rhetoric of transnational institutions that have entered the fray to provide the knowledge, technology, and skills necessary to manage Delhi's water sector. These discourses of power were built around the framework of the so-called "public–private partnership" in water reform policy. The projects of 24/7 Water Supply and the Sonia Vihar Plant were initiated to address the problems of Delhi's water supply (e.g. intermittent and insufficient water supply, bad quality, availability, etc.) and aimed to improve the potability of the water, build the water supply infrastructure, and provide social and economic development as well as environmental sustainability. The water reform discourses incorporated these claims, and solutions were presented in the politically neutral terms of a technical frame.

The claims of the Government of Delhi in the project represented water in a broad, macroeconomic and technical frame of "managing" development. The consensus on policy reform can be associated with a broader agreement between the donor communities, central Government, and the state of Delhi. These relationships between donors, lenders, and the Governments are the factors that comprise the DWSSSRP, which represents a policy instrument that is constructed as "state owned" and relies on a foundation of consultation between the Government of Delhi and civil society concerning water reform. It is perceived as offering an opportunity for a range of actors to engage legitimately in policy formulation. Simultaneously however, it remains an instrument of political and economic conditionality which forms an essential part of the narrative established by international financial institutions. As the project unfolded, an unprecedented range of policy spaces emerged, but the potential of these spaces depended both on the terms of invitation to invited stakeholders and on the collective action of uninvited stakeholders through the formation of social movements that frequently fall outside the bounds of the current hegemonic discourse of water reform. These uninvited stakeholders became an important entry point to the understanding of the dynamics and development of alternative discourses in the agenda of water policy reform. The next chapter discusses the emergence of these spaces and the methods and counterclaims deployed by these groups to create opportunities for policy change.

Notes

1 The World Bank's intention with its participatory approach is reflected in its refusal to participate in the public hearing held in Delhi on 17 October 2005 to discuss the privatization of the water sector in Delhi and the 24/7 project. Country Director Carter invited Aruna Roy, the social activist and Magsaysay Award winner, to discuss the

Delhi Water Supply and Sewerage Reform Project in his office over a "cup of tea." She turned down this offer on the grounds that discussions on the project needed to be taken up with the Delhi citizens so that people could play a more basic role in understanding and choosing policy options (Mehdudia 2005b).

2 What the Government of Delhi did not disclose is that the bulk water supply for treatment at the input zone of the treatment plant and at the distribution zone point for the households is the responsibility of the DJB. If the DJB does not provide the water, the company is absolved of all contractual obligations. In fact, the DJB will have to pay a penalty to the plant for not supplying water on time (see Chapter 5). Similarly, the company is absolved of the 24/7 vision for households. The company's responsibility rests only on providing water from the treatment plant to the distribution point from where water will be distributed to households by the DJB.

3 There is no doubt that "price" is an issue that needs to be addressed in the Indian context, and there is truth in the fact that subsidies actually do not reach the poor, but the solution offered in terms of private being effective and public being ineffective does not reflect the Indian understanding of sociopolitical realities.

4 Presently, the people in the JJ clusters get water from stand posts, water tankers, leaking pipes and tube wells. They are not a part of the formal water distribution network because they do not have tenured land rights. They therefore either depend on free water or illegal water from leaking pipes. The idea of group meters increases the monetary burden on people who live on less than a dollar a day. Groups meter connections require monthly payments and with no subsidies and no process to recover the dues for used water in a household from a shared meter would lead to conflicts among the neighbors sharing the meter (RFSTE 2006).

7   Creating spaces for change

Collective action on the water reform project

Chapter 2 argued that policy processes are a complex configuration of actors and interests that have different opportunities for influencing and shaping policy processes as they position themselves in the different sites in which policy making takes place. In this chapter, particular attention is paid to the opportunities that the emergence of new "policy spaces" offer for the engagement of those conventionally excluded from policy deliberations. As discussed in Chapter 2, policy spaces are moments of intervention or events that bring new opportunities to reconfigure relationships between actors and open up possibilities for a shift in direction. An examination of policy spaces gives us a better understanding of the dynamics of power, agency, and knowledge in shaping water policy. Moving away from the traditional view of policy space that tended to include the various departments of government, this chapter speaks to two broad clusters of policy spaces – invited spaces from above created by institutions like the state or donors and spaces created from below through independent forms of resistance and social action. An analysis of these actors provides an important entry point to understanding the dynamics and development of alternative narratives from the mainstream agenda of the urban water reform project. The following sections pay particular attention to opportunities that emerged as "policy spaces" for those conventionally marginalized from policy deliberations.

Invited spaces in the policy process

Mainstream narratives in the Government of Delhi's water reform project report (DJB 2004: 8) mention the involvement of all stakeholders in the policy change initiative. The report claims that the success of the project would depend on addressing the concerns of these people. Two workshops were held in March and May of 2004 to determine the needs and priorities of these people. However, the use of these consultative processes and the ambiguity of these moves are evident in statements that alternate between celebrating them as opportunities for participation and democracy (Schonwalder 1997: 261) and analyzing their deployment by powerful actors to dissipate or subsume resistance (Kanyinga 1998), with the strategic significance of presenting the decision as uncontroversial and based on consensus.

In practice, as was revealed by some groups that were invited, "consultation" was the dominant mode so that ideas of civil society and stakeholders were solicited but the draft documentation was already drawn up and remained unaltered in the consultation process. These experiences, according to Collins (2000), describe the boundaries of the space offered, and the key elements of the narratives of these stakeholders remain constrained. The stakeholders or the nongovernmental organizations (NGOs) have fairly limited opportunities to influence the framing of policy in invited spaces.

Collective action as policy space

The other means of bringing in new voices and discourses in the policy process is through spaces from below. This section explores some of the ways in which civil society actors articulated alternative views on water reform policy beyond state invited spaces. In the Delhi case study, given the fact that policy formulation was conducted in secrecy and the consultative processes had limited effects in influencing policy, the articulation of alternative agendas took place through the medium of two alliances and coalitions – with NGOs at their centers – which acted to open up new spaces or occupy existing ones in their efforts to change policy. These were the Water Liberation Campaign (WLC) and the Citizens Front for Water Democracy (CFWD).

The formation of these alliances occurred around issues that are intimately related to a multidimensional understanding of water. Despite this, the way that these social movements framed and acted upon water policy reform issues frequently fell outside the bounds of the current hegemonic discourse of water reforms in Delhi. As such these movements and alliances are an important means of understanding the dynamics and development of alternatives narratives about water. Most research on social movements emphasizes that marginalized social groups make claims on authorities based on their own interpretation of dominant ideologies (Houtzanger 1999). By making claims around rights and justice issues that have not been delivered by the state these movements seek to politicize their actions. This intersection between these movements and the state becomes an important space for discursive engagement and action.

In the case of Delhi water policy reform, social movements became an important mechanism for the creation of spaces for alternative visions of water resource development and management. The processes behind the building of the Sonia Vihar water treatment plant were completed in secrecy and it was only when reports about the contract awarded to Ondeo Degremont on a ten-year Build-Operate-Transfer (BOT) basis appeared in the press that opposition groups composed of concerned citizens in Delhi and along the banks of the Ganga responded to this development and in doing so helped to frame the issues of water reform in terms of a struggle against privatization. Each of the issues these groups brought to the fore became a key element in the resistance to water privatization. The fact that most of these people had been unaware of the Government of Delhi's plan for reforming urban water supply under a BOT project added fuel to their indignation. Additional concerns were raised in response to an increasing number of newspaper reports detailing the role of Ondeo Degremont and the 24/7 project in Delhi. Ganga Yatras, or river pilgrimages, were organized annually between 2002 and 2004 to rejuvenate the living culture of the sacred Ganga and the significance of water to the Hindus. A Jal Swaraj Yatra (journey in support of water democracy) was organized from 15–22 March 2004 on World Water Day (RFSTE 2005). A million people were involved and 150,000 signed a hundred-meter "river" of cloth to protest privatization of water in Delhi (Sehgal 2007).

In September 2005, more than 20 eminent citizens wrote to Sonia Gandhi, the Chairman of the National Advisory Council (NAC), expressing serious reservations about the water reform project. The signatories included Muchkund Dubey (former Foreign Secretary), Bipin Chandra (eminent historian), Romila Thapar (eminent historian), Deepak Nayar (former Vice Chancellor of Delhi University), Padma Bhushan Awardee G. P. Talwar, G. S. Bhalla (former member of the Planning Commission), Chitra Mudgal (former chairperson of Prasar Bharti), Prabhat Joshi (eminent journalist) and Aruna Roy (member of the NAC). The protestors challenged the claims made by the Government of Delhi for the Sonia Vihar Plant and the 24/7 project, citing the Delhi Water Supply and Sewerage Policy documents which were based on the reports of consultants PricewaterhouseCoopers (PwC) and which exposed the involvement of the World Bank. Stating that the proposed round-the-clock supply in South Delhi was nothing but a complete sellout to multinational water companies, NGOs and Resident Welfare Associations (RWAs) across the capital geared up against the Government's alleged move to privatize water distribution (*The Hindu* 12 July 2005). In large part, opposition arose because of doubts about whether private markets could address the many different social, environmental and political aspects of water in the territory. These alliances became the policy spaces which began as moments of intervention or events that aimed at creating new opportunities to reconfigure relationships between actors and produce a shift in policy production. Thus, in the efforts of the WLC and CFWD to protest the Government's policy decision over water privatization and open up the policy space, several closely linked and interwoven themes to support their positions emerged. The activities of these groups and alliances demonstrated the process of a movement which emerged from local action spaces, and which addressed its campaign both at the state and at the multinational corporations (MNCs). These groups used mass mobilization and direct confrontation to construct a unified response against the state, making counterclaims against the Government's justification of the Delhi water reform project.

Activist claims against the project: creating opportunities for change

Water is a commons and a basic human right

Claims about water as a human or public right formed a significant segment of the claims against the privatization of water reported in a variety of newspapers and repeated over the length of the controversy. Specifically, these claims argued that "water is fundamental to existence" and therefore "forced scarcity is a violation of the fundamental right to life" (*The Hindu* 19 December 2003). Newspapers repeatedly pointed out that it was a governmental duty – specified in the constitution – to ensure water rights, and that the poor in particular are dependent on the Government for its continued provision. The activist groups also pointed out that the right to water is common to all beings and is a gift of creation. It is a natural, "common" right:.

> Common Rights go hand in hand with common responsibility – a common responsibility to conserve and share water, use it sustainably, and share it equitably. Blind commitment to privatization and attack on community rights and commons amounts to hydro-apartheid.
>
> (Shiva 2003)

Claims made by groups like the WLC and CWD referred to the belief that water rights did not originate with the state; they evolved out of a given ecological context of human existence – that people have the right to use water and Government and corporations cannot usurp this right. The campaigners also emphasized that water as a fundamental human right has been implicitly supported by international law, declarations, and state practice.[1] In addition, given the fact that only a third of Indians have adequate access to safe water, the activists argued that the right to water requires Governments to progressively increase the number of people with safe, affordable, and convenient access to drinking water. The WLC claimed that the right to water also includes obligations for nondiscriminatory access to water, especially by the marginalized and vulnerable sections of society. In the case of the Delhi water reform project, all these conditions would not be met if water distribution was transferred to private players as proposed. Privatization, activists said, was not a solution to the water crisis. "Privatization is the enclosure of the water commons" (Shiva 2005).[2] Furthermore, according to the activists, water resources flow from nature and not from the rules of the market. "The only long-term and prudent water policy is to recognize nature's limits, live within the water cycle, and guarantee every Indian their fundamental right to water" (Shiva 2005).

The campaigners claimed that in a neoliberalizing state like India, attempts at commons-use management today are mistakenly decried as obstacles to sustainability, progress, and growth. Community rights have long been a part of the cultural traditions of India, and even the colonial rulers did not hand over water systems to private corporations. The protestors argued that water is the source of life and belongs to all the inhabitants of the earth in common. No one either individually or as a group can be allowed the right to make water into a private property. Water, opposition groups argued, "is a commons, a public good," rather than a source of private wealth and profit for individuals.

While many intellectuals within the campaign agreed that there are issues of pricing and efficiency that need to be addressed, they insisted that water should not be described as a commodity. "Water is not a commodity to be bought and sold for profit, but rather a common good. It is a public utility that can be priced but within certain socio-economic parameters" (Viswanathan 2003). In a personal interview conducted on 4 December 2004, Dr Vandana Shiva, Convener of the CFWD and Director of the RFSTE, reiterated the sacred common heritage that human beings share in water: "In fact in most societies private ownership of water has been prohibited. Water is a moving, wandering thing, and must of necessity continue to be common by law of nature so that it can only have temporary, transient property rights therein."

The economic valuation of water, activists argued, tends to disregard the human right to water claimed by citizens, which are the ultimate owner of the resource. Since water cannot be exchanged in an open market the practices of free market economics do not apply to the provision of this resource in India in general, and in Delhi in particular, where a third of the population lives without piped connections. Based on these perceptions of water, the WLC and the CFWD claimed that the ongoing urban water reform project was not about public–private partnership (PPP) as the Government of Delhi consistently reiterated in all policy documents, newspaper reports, and interviews. It was the road to ultimate privatization of the city's water supply.

Project outcome: no expertise, technology and cost recovery

One of the main arguments made by the WLC and the CFWD was aimed at the workability and the feasibility of the water reform project as claimed by the World Bank, consultants, and the Delhi Water Supply and Sewerage Policy document. The campaigners challenged the consultants' claim that lack of experience among Indians in providing treated water 24/7 made it imperative to have outsourcing and private management of the two pilot operation zones (South I and South II) for a period of five years. Opponents of the scheme were united in the view that the private sector had no monopoly on expertise. They believed that sufficient expertise is available in the public sector, and that in fact any private players brought in from outside would lack ground experience of water distribution in India because they would be unaware of Indian social realities.

The CFWD also argued that these international private players would not bring in any new technology. "We do not have a dearth of local capital in water treatment and supply," said Commander Sinha of Paani Morcha, a member NGO of the WLC, in a personal interview conducted on 10 September 2004, "but international companies do better marketing to specific individuals handling these projects." Dr Shiva of the WLC and the CFWD concurred in a personal interview: "These companies bring no new technology. They just harvest it – to collect twice the amount to build the plant and then take ten times the amount to run the water system. They don't bring us any investment but take a share in our borrowing in the name of technology and quality."

The members of the WLC and the CFWD also quoted newspaper reports that raised claims about the treated water quality of the French MNC Suez in India and in other parts of the world. These reports suggested that the quality and service became abysmal as the French company began to focus solely on profit. For example, Suez was reportedly given a 20-year, $428 million contract in Atlanta after paying for a $12,000 trip to Paris for the mayor. After the changeover, residents complained of foul-tasting, brown-colored water. In Milwaukee, an audit found that a Suez subsidiary shut down pumps during hours of peak electricity demand to save money, creating overflows that sent raw sewage into Lake Michigan and rivers in the area. Another French company, Veolia, was found to have dumped raw sewage into the Mississippi River. The same company got a contract in Indianapolis, where complaints about water quality tripled after it took over (*Newstrack* 29 May 2006). These articles argued that where profit becomes the primary driver of the process, issues of quality and service take a back seat.

Through letters to the Prime Minister and Sonia Gandhi, Chair of the National Advisory Council, intellectuals and activists brought to the notice of the Government of Delhi that in several other parts of the world water quality had deteriorated, tariffs had skyrocketed and riots had broken out. The companies had had to flee some countries and made profits even when they left the country by filing multimillion dollar suits against governments. The activists argued that these projects that run in the name of PPPs mainly benefitted the MNCs.

Many intellectuals, however, agreed that high costs and low revenues had made the Delhi Jal Board (DJB) bankrupt. It was distributing only 60 percent of the water it produced and generated revenue from about 40 percent. Only about 30 percent of the water it produced generated cash revenues as compared to 85 percent from well-functioning water utilities. Cost recovery then becomes an issue and tariff hikes can curb wasteful consumption of water and reduce the quantum of nonrevenue water. But, as the intellectuals argued, "Can't the DJB adopt these measures indigenously?" (Sharma, Personal Interview 7 September 2004). The companies charge enormous sums as performance, management, and maintenance fees while the onus of providing raw water and water to the household would rest with the DJB. This is a story, they said, that has a script: "We pocket the profits; the losses are your liability" (ibid).

Delhi water project to benefit multinationals

The main argument that NGOs and activists offered in opposition to this project was that MNCs are profit driven and that any water projects run will primarily benefit MNCs. "Multinationals and Profit Sharks versus the People of India," *The Economist* summed up the controversy surrounding water in the DJB in India in 2005. Most newspaper reports stressed that the private sector is in business for profit and that remains its primary motive. Privatization was handing the resources of the country to corporations and MNCs. The distinguished author and leading activist, Arundhati Roy, writing in *The Hindu* (18 October 2005) developed the argument further:

> History proves that companies are only concerned about their profits and least bothered about people's welfare. Privatization of water will not be a

good experience for Delhi, as it has been proved in similar cases in other cities the world over.

Quoting a variety of sources,[3] newspapers reflected the general view of the populace that MNCs like Suez enter India to reap unjustifiably high profits. The reality of the marketplace, they argued, is that private sector financiers have such bargaining clout that they can dictate the terms of BOT projects like the Sonia Vihar Plant. Most importantly, they claimed that in the BOT projects, the Government of Delhi bears the demand and market risks and provides a guaranteed return to the investors, either by providing a subsidy when needed or by incorporating the return component in the tariffs. Specifically, local newspapers reported the ire of the WLC directed against Suez, which they believed was the intended beneficiary of the reform projects in Delhi. The WLC's slogan, as reported in a variety of sources was, "Suez pay the full costs, or quit India" (*The Hindu* 31 August 2003; Shiva and Jalees 2003; Personal Interviews 2004).

The CFWD and the WLC argued that the BOT agreement with Ondeo Degremont was totally biased against the public and unfairly privileged the MNC. In a letter to the chief executive officer (CEO) of the DJB, Rakesh Mohan, the Convenor of the CFWD quoted clause 3.6 of the agreement that obliged the DJB to provide "free of charge raw water." The activists claimed that this "raw water" was coming from the sacred Ganga and was to be brought from the Tehri Dam. In the letter Dr Shiva (5 April 2005) claimed that public investment made to bring the Ganga water to the Sonia Vihar Plant included:

> Rs. 10,000 crores for constructing the dam, Rs. 147,453 crores for the costs of construction of the Upper Ganga Canal and Rs. 111.31 crores for laying pipelines from Muradnagar to Sonia Vihar – i.e. a total of Rs. 158,149.31 crores of public investment. While this is the cost of water, Suez gets it all for free.

Clause 4.21 of the agreement, activists said, allows Suez to use free electricity to be supplied by the DJB. With privatization of power already in place, activists claimed that this would put yet another burden on the people. Activists also alleged that the plant could have been constructed for Rs. 100 crores – Ondeo Degremont was awarded 189.90 crores to build it. This, they claimed, was a waste of precious public wealth and resulted in a tenfold hike in water tariffs to meet the budget deficit of the DJB (Personal Interviews with members of the WLC in 2005).

News reports consistently claimed that the agreement imposes stiff penalties on the DJB and offers Ondeo Degremont easy incentives. Activists pointed out that what appears to be an unusually fair agreement by World Bank standards actually comes with a catch. The contract states that a fixed management fee will be paid to the company and if it fails to meet its target it will have to compensate the DJB and its consumers in the form of penalties and lost bonuses. However:

> The DJB must insure that the plant receives an adequate supply of water, in the absence of which it is liable to pay for base service charges and inventory charges for consumables and chemicals. Estimates for such penalties range between Rs. 50,000 to 80,000 a day.
>
> (Sethi 2005)[4]

Newspaper reports and activists argued that the contract provided for incentives for overproduction and energy savings. While performance based incentives make sense in principle, the parameters set for Ondeo Degremont are far below those set for other DJB plants. The contract is designed in such a way as to ensure that the company always meets its targets:

> Degremont has been allowed 232 Kwhr [kilowatt hour] per million liters of water treated while DJB plants of similar capacity consume between 170–80 Kwhr per million liters of water.
>
> (Sethi 2005)

Protestors alleged that public money was being diverted to global corporations while the poor faced the tariffs hike and the middle classes would pay for the profits guaranteed to Suez under a fixed operation and maintenance fee of Rs. 2.80 crores a month. "All infrastructure for the plant – from pipelines to canals – has already been financed by the Indian Government. Why do we need Suez if it concerns our money, our resources and our people? (Personal Interview 11 September 2004).

Thus, activists argued, the responsibility of water supply, both in the treatment plant and in the 24/7 project, rested with the DJB. The NGO Parivartan claimed that if the DJB failed to maintain its end of the contract, the company would become free of its contractual obligations. Essentially, the contract required every operating zone to be divided into several district-metering areas (DMAs). As long as the company provided water at the input point, it is deemed to have done its duty and can claim incentives. The DJB must provide water from DMAs to every household 24/7. Thus, the entire project revolved around paying private companies to distribute water from the operating zone input to the DMA input.

A water company would be paid about 5 crores per annum (50 million dollars) fixed fee to manage a zone, which would go to meet the salaries of four expert consultants for each zone at the rate of $24,000 per month. On the all-Delhi level it comes to an additional expenditure of Rs. 105 crores per annum of 21 zones for employing 84 experts. This amounts to about 20% of the total O&M budget of the DJB. This project is a financial bonanza for water companies.

(Parivartan 2004)

Another argument that supported the activists' claim involved Ondeo Degremont's capital investments to maintain the infrastructure. Activists allege that all these expenses are a series of blank checks being written by the Government of Delhi, as there is no categorical explanation of any checks and balances on expenditure by these companies. Thus, there were no provisions for the company's accountability. As noted by Parivartan (2004):

It seems that the company will present wish lists and the government will simply write the checks. Since the company is not spending any money, it will have little incentive to show prudence in expenditure. Costs of operations and costs of implementing capital works would go up substantially.

Parivartan (2004)

The Government of Delhi committed itself to making huge investments to the tune of Rs. 3,500 crores during the Tenth Five-Year Plan, the burden of which would ultimately fall on consumers (*The Hindu* 13 July 2005). Summing up their claims that the Government's project benefitted the corporation, critics contended that the role of the MNCs is "not to bring you any investment but take a share in our borrowing. That is why we have started a full cost recovery campaign from Suez for the public money they are using to make profits" (Shiva Personal Interview 4 December 2004).

Arvind Kejriwal of Parivartan (2002) commented that "The Project is likely to have an irreversible impact on the water and sewerage sector in Delhi .... It will neither lead to 24/7 water supply nor any improvement in the performance of DJB .... The project just ensures a financial bonanza for the multinational water companies, which would be paid by the consumers in the form of higher tariffs." Many in the WLC also believed that if the Government of Delhi handed over the DJB's water distribution function to MNCs, it would not find a solution to the DJB's institutional malaise but merely put it in the background – behind an expensive and risky management contract that benefits the company.

Project promoted the global practices of the World Bank

Activists also alluded to the role of the World Bank in designing these projects as its current development agenda advocates the management of water resources by private players. There seemed to be an implicit understanding on the role of the World Bank, especially where terms and conditions of the tenders were framed in such a manner that no Indian company was likely to qualify for the tendering process. These conditions substantiate the claims of the protesters that the World Bank's loan policy was helping to turn natural resources like water into commodities and thus making them scarce for the poor. Most newspapers repeatedly echoed the campaigners' claim that the World Bank policy favors the involvement of private players in urban water supplies and managements. Reports from various sources considered the hegemonistic designs of the World Bank and MNC culture under this new water policy as neoimperialism in its worst manifestation.

Most activists and protestors felt that the shifts in the DJB's policies were directly linked to the World Trade Organization (WTO), the General Agreement on Tariffs and Trade (GATT), and the World Bank. They claimed that a variety of moves have been quietly under way for some time with the active prompting of the World Bank, the IMF and the Asian Development Bank (ADB) to privatize water supply utilities in India. Specifically, in the case of Delhi, Vandana Shiva of Navdanya/RFSTE and the main leader of the WLC, alluded to the water project's global connection:

> The World Bank has written clearly on sector adjustment roles. The entire policy document for Delhi has been prepared by World Bank consultants Price Waterhouse. Even now every step in the planning and policy decision is coming out of the World Bank. So every time someone asks why the Government of Delhi decided this? I say the Government of Delhi does not decide. The World Bank decides, and decision makers who put pressure on the World Bank too are moving the World Bank money. So in a way the World Bank is also being used by giant global water corporations who have a lot to gain by these shifts.
>
> (Personal Interview 4 December 2004)

Delhi's water project was based entirely on the reports of PwC and GKW, which contained World Bank jargon phrases such as "subsidy transparency," "full cost recovery," "efficiency," "depoliticization of tariff," etc. – World Bank terminology for the corporatization and commoditization of public services and utilities appeared frequently, as did references to the phasing out of subsidies and cross subsidies (Sethi 2005). The protestors alleged that the World Bank's rhetoric of PPPs was being used for the Sonia Vihar Plant and the 24/7 project but, they stated, semantics did not change the reality of the privatization objectives of the World Bank.

Project is anti-poor

The Government of Delhi's claim that private players will specifically benefit the poor and that the urban water supply reform process was essentially targeted to the poor was challenged in a number of newspaper reports and activist interviews. Newspapers consistently reported claims that the role of private players in urban water supply would hurt the poorer section of the society. For example, *The Hindu* (21 April 2005) reported the assertion of the activists' claims that the Government of Delhi was "ignoring the interests of the poor and instead catering to the interests of the multinational companies and business houses." *The Financial Express* (22 March 2005) reported that "The current push to privatize water is a recipe for destroying our scarce water resources and for excluding the poor from their water share."

The CFWD argued that the poor in Delhi are not part of a formal distribution network because they do not have land tenure rights. As a result, they have to depend upon either "free water" from (stand posts and tankers) or "illegal water" (from leaking pipes). These sources of water are treated by the Government of Delhi as part of nonrevenue water. Since both the consultants –PwC and GkW – and the DJB's *Delhi Water Supply and Sewerage Sector Reform Project* aimed at reducing nonrevenue water, activists alleged that all these sources would be shut and there would be no free water. Water tankers, tube wells, and public taps in poor settlements would first be cut off and then five poor families would supposedly be provided one group connection in the Jhuggi Jhopri (JJ) clusters. While they could pull separate taps to their houses, activists said they would have one meter and would be collectively responsible for paying the bill. There is, however, no proposal for investment on extending the water distribution network to JJ clusters. How then, opponents argued, would the group connections be provided and how would this lead to targeted interventions for the poor in accessing water? In any event, group meter connections, they contend, would only lead to conflict amongst the users (RFSTE 2006).

The WLC and CFWD also claimed that since much of the JJ cluster population lived on less than a dollar a day they would not have any money to get the metered connection established in the first place, which implies that most of them would be denied access to water. Private operators will supposedly provide 24/7 supply to all by plugging all leaks, but activists said that loss reduction targets might be rendered meaningless by the absence of adequate information to set a realistic baseline. They planned to replace "informal" sources of supply (on which a third of Delhi relies) with piped connections, but with what obligation and at what cost? (RFSTE 2006).

Given the highly inequitable distribution of income and resources in India, reports mentioned that it is impossible for most poor people to pay the high charges that inevitably accompany privatization. Newspaper reports noted that the Government of Delhi did not talk about low cost supply to the poor but only "urban poor paying less for safe, piped water than what they do through illegal and informal vendors." In reality, according to opponents of the project, most of the poor would end up paying much more than they were already paying. It should also be noted that the ADB and the World Bank strongly pushed for the withdrawal of both cross and direct subsidies in the water supply sector – the only mechanisms that can ensure low cost supply to the poor (Dharmadhikary 2003; Parivartan meeting on World Bank policies 23 March 2004).

Most activists of the WLC and CFWD believed that "fraudulent steps" toward privatization of the DJB would lead to encroachment on the rights of the poor people who would be forced to spend more under the World Bank agenda. Reports consistently claimed that the project would adversely affect the urban poor and slum dwellers that may not have the money to pay to access drinking water (David 2005). They argued that in a one sided negotiation process, weighing heavily on the side of sponsors with their expertise and influence, the balance of risk allocation is bound to favor private operators. The poor, as the least attractive customers, would be likely bear the brunt of these risks.

Another argument made by the protestors was that water has serious health and environmental externalities – the price of neglect could be life-threatening. This was a worry in poorer areas, which would likely be low on the list of private operators' priorities (consulting reports indicated that they expected to rely on relationships with NGOs to serve these areas). These conditions might compel them to look for dirtier and more polluted sources.

These campaigners therefore challenged the Government of Delhi's "targeted interventions for the poor." According to them the project could only benefit the poor if they had a voice in both policy formulation and grievance redress. Further, campaigners argued, private operators have a greater incentive to divert water from the poor to lucrative customers:

If the project delivers water 24 hours a day, it will only be for the privileged few at the cost of the poor because the Bank believes in providing "all water for some and no water for most, whereas we believe in some for all." The Government should ensure that water is not just for the rich but also for poorer sections of society.

(Rajendra Singh Personal Interview 9 December 2004)

According to the WLC and CFWD, the hike in tariffs that preceded the privatization process was another signal that poor consumers were going to be the worst hit in this process of water policy reform. The consultants, they claimed, suggested phasing out subsidies and cross subsidies and setting up a water regulator to determine tariffs on the principles of full cost recovery. According to the WWA, if the suggestions made in the reports submitted by the consultants were eventually implemented, the monthly water bill of an average middle class family in the capital who was paying Rs. 192 would go up immediately to Rs. 990 (calculated at the present level of operating expenses), and that of a family living in a slum cluster who was paying Rs. 52 would increase to Rs. 200 (RFSTE 2006). While operating expenses were likely to increase sharply under private water companies, the actual rise in tariffs would be far higher (Parivartan 2004). Such a process was bound to adversely affect the poor, especially women.

Activists further stressed that the issue of water needs to be addressed as a gender issue as well as a political issue. If water policy reforms were accomplished as planned, women would be the most affected in Delhi. A survey done by the New Delhi Research Foundation for Science, Technology and Ecology (RFSTE) showed that most women collect water for the family from standpipes and other nearby sources, by standing in long lines for hours at a community tap or by collecting from leakages in the public utility pipelines. Most respondents were against installing meters as their husbands hardly earned enough to make a living and would not be in a position to pay for water (RFSTE 2006). Thus the WLC and the CFWD believed that both the poor and women would be adversely affected by the project.

The WLC and CFWD alleged that while the rhetoric for tariff restructuring appears to be pro-poor, it was in fact anti-poor because it seeks to transform a public good to a privatized commodity to which access would be proportional to the ability to pay. The CFWD quotes the DJB report as stating:

DJB shall strive to undertake metered supplies in all such colonies (JJ clusters) and rural areas. However, till such time that this is achieved, the colonies and rural area consumers shall pay the access charge fixed for domestic consumption and an additional amount of volumetric consumption, based on an assumption of 10KL consumption.

(RFSTE 2006: 24)

"In other words," the CFWD comments, "the poor will be paying Rs. 40 – even without access to water" (RFSTE 2006: 24). Further, activists argued, the poor will very likely see higher prices and reduced access and yet the scope of the reforms does not seem to address critical problems of inequitable distribution and affordability.

Project creates inequity

For the campaigners, at the heart of the struggle over Delhi's water reform project lay not only a human rights case in favor of universal access to water, but also a debate about core governmental values of equity.[5] "Despite the DJB's claim of equal allocation of water, the supply of drinking water in the capital is characterized by vastly unequal distribution, with posh colonies and VIP areas getting several times more than the supply given to rural areas and resettlement colonies" (CSE 2003). The reform project would, they argued, make its supply worse:

> Two thirds of the city's population gets only 5% of the 3,600 million liters that are officially supplied, whereas the rich and privileged in Delhi get a staggering 400–500 liters per capita every day. There are some 1,600 unauthorized colonies and 1,100 slums waiting to get tap water. Not only have the rich been the prime beneficiaries of subsidized water treatment plant at Nangloi which runs at its half capacity, the Sonia Vihar plant again will supply water to the posh South Delhi colonies and increase this already existing inequity.
>
> (Suresh Babu Personal Interview 11December 2004)

Activists also claimed that private operators would have greater incentive to divert water from the poor to the lucrative communities. These actions, which would give private operators a free hand, protestors believed, infringed on fundamental and democratic rights.

Project is against democratic rights

Proposed water sector reforms in Delhi were considered nothing short of a disaster by eminent social worker and Magasaya Award winner Aruna Roy. Roy argued that "the technical reasons cannot be more important than the ethical and democratic issues" (Mehdudia 2005a). Elaborating on her statement, Roy asserted that the terms and conditions of the donor agency, consultants, and MNC pressures that had dominated the reform process over the last several years had seldom been discussed and shared with the people, in contravention of the spirit of democratic and transparent governance.

Many intellectuals, in their 2 September 2005 letter to Sonia Gandhi, the Chair of the National Advisory Council for Delhi, stated that the project, which would influence the lives of millions of people in Delhi, gave the World Bank the last say in all matters to the total exclusion of Government of Delhi, the elected representatives, and the people. The terms and conditions appeared to grossly undermine all norms of democratic decision making in India.

Privatization of something as basic as water, which one needs to survive, is clearly violating the basic principles of democracy. The WLC argued that private interest groups systematically ignore the option of community control over water. The New Delhi based RFSTE and other NGOs criticized the project as:

> a direct attack on the country's water democracy, water culture, people's right to water and sustainability of water resources. The government should enter a public–public partnership with the people rather than a private–public partnership. The involvement of citizens in all such situations is very important and we are competent to handle our own valuable resources rather than seek the services of global multinational giants to set things right for us.
>
> (Shiva Press conference 25 October 2005)

Both the WLC and the CFWD felt that water is such a basic need that it should remain in the hands of the public and that water democracy, not water privatization, would lead to the sustainable and equitable use of water. The general feeling among these groups and alliances was that corporate control of water was facilitated by the creation of corporate states – states that centralize power, destroy federal structures and the constitutional fabric, and usurp and erode fundamental community rights. The struggle for water democracy was also their struggle against the state of Delhi because they felt that the market rules through coercive, anti-people, undemocratic states. Privatization of water denies local communities their water rights and access to water. Presenting an alternative to Delhi's water project, the CFWD argued:

> We do not need privatization or river diversions to address Delhi's water problems. We have shown how with equitable distribution and a combination

of conservation, recycling, and reduction in use, Delhi's water needs can be met locally. We need democracy and conservation. The seeds for the water democracy movement in Delhi have been sown. We now have to nurture them to reclaim water as a commons and a public good.

(Declaration of the CFWD 22 March 2004)

Collectively all water activists took a pledge to create water democracy (Jal Swaraj) and defend the people's right to water as a commons. They view these rights as being infringed upon and usurped by the World Bank and private corporations. Thus the contract to Suez, the consultancy work being awarded to PwC at the behest of the World Bank, and the rise in water tariffs were cited as examples of the subversion of democracy.

The memorandum that a delegation from CFWD handed to the World Bank President Paul Wolfowitz during his visit (RFSTE 2005) accused the World Bank of a water apartheid policy, and called their struggle with the World Bank "an issue of ownership of our rights, our participation in decision making and commitment to water democracy." The Delhi water project, the activists argued, not only created issues of subsistence, affordability, inequity, and subversion of democracy, but was also a direct attack on the country's cultural traditions and conventions that have been followed since the times of the Vedic civilization, which preceded the centuries of India's growth and modernization.

Cultural and religious claims to water

Groups protesting the Government of Delhi's PPP, which they claimed to be "back door privatization," used cultural and religious claims to support their struggle. Water and religion, they argued, are inextricably woven into the pattern of Indian life. From the birth of a child, to pre-marriage rituals and death, water has deep spiritual and cultural meanings. Prayers are offered to the water and rain gods in the Hindu panthenon. If the rivers are privatized and water diverted for profit at the behest of thirsty corporations, the WLC felt it would leave the Indian culture in ruins. "This is something we cannot allow" (Personal Interview WLC member 11 November 2004).

So on the banks of the Ganga River on 9 August 2002, the anniversary of the Quit India Movement for independence, a group of citizen campaigners took a solemn pledge to protect its sacred waters:

We will never let the river be sold to any multinational corporations. Ganga is revered as a mother (Ganga Maa) and prayed to and on its banks important ceremonies starting from birth to death are performed (according to Hindu religious practices). We will not allow our mother or its water to be sold to Suez Degremont or any other corporations. The sacred waters of the Ganga cannot be the property of any one individual or a company. Our mother Ganga is not for sale. We boycott the commodification and privatization of the Ganga and any other water resources.

(Hardwar Declaration 2002)

This pledge, known as the Hardwar Declaration, staked out the position of grass-roots activists on the urban water policy reform in the National Capital Territory (NCT) of Delhi.

The Indian water traditions were another argument on which the activists staked their claims. When guests arrive at someone's home in India, the first thing they are offered is water. So water, the WLC says, is an important part of Indian culture. Indian civilization developed around rivers and rivers, sacred to Indians, have a very special role to play in Indian society and culture. Because water is considered a sacred gift and a creative force, essential for all life, it occupies a central place in the Hindu religion and belief system because rivers are considered purifiers of sin and bridges between the human and the divine. Dr Shiva and Commander Sinha of Paani Morcha, a member of the WLC, emphasized that water in India has a high cultural and religious value, which is priceless: "We worship our rivers and place a high spiritual and ecological value on them. Water is sacred in India and cannot be allowed to be commoditized" (Sinha Personal Interview 3 December 2004).

Thus the Ganga, whose water was diverted by Suez, became the symbol of the campaigners' struggle against privatization.

A river like the Ganga is my mother and I cannot allow my mother to be sold to a multinational. To put the fate of the most sacred river and beliefs of millions of people in the hands of such a corrupt transnational corporation is an assault on the culture and foundations of the people of India

(Shiva Personal Interview 4 December 2004)

Strangers the day before, dozens of voluntary groups gathered in the capital for a common cause. Their ire was particularly directed against MNCs involved in the water treatment and supply systems as well as bottled water plants.

Condemning the tariff hike by the Government of Delhi on 30 November 2004, the CFWD and the WLC said that this hike was also an attack on the tra-ditional Indian values attached to water. That is why campaigners were outraged when destitute homes, disabled homes, orphanages, religious premises, charitable guest houses (Dharamshalas), cremation grounds – which had been provided free or subsidized – were brought into the sphere of taxable entities. The most appalling example, the CFWD claimed, was the Piaos.[6] In the Hindu tradition, the campaigners pointed out, it is believed that if you provide free water to a thirsty person your virtuous deed will get you credit on your balance sheet in Heaven. Even the colonial state did not charge any taxes to the Piaos during their rule, the campaigners asserted, but the post-colonial state demands that the Piaos have to buy water to give water to the thirsty, thus making the traditional water temples a water market.

Commenting on the post-colonial state that devalues our traditions and culture through such privatization practices, Rajendra Singh, popularly known as "Water Man," remarked:

> They [the post-colonial state] are worse than the colonial state. At least the British respected our sentiments of religion and culture. They used to arrange for the Piaos themselves in the summer free of cost for people to stop by and drink water in India's tropical climate. The new state does not exhibit the colonial tendencies openly but they are more dangerous because they say the shift in policy is for the benefit of the society yet they are exploiting us, being pawns in the hands of the global players.
>
> (Singh Personal Interview 7 December 2004)

Alternatives to public–private partnerships

The WLC and the CFWD claimed that the Government of Delhi was being seduced into hastily applying borrowed models with track records of failure. At the very least, they believed, the Government should provide full disclosure of the proposed reforms, including explanations of how risks of nonperformance and cost padding would be mitigated. They also believed that the public must get involved to demand transparency, evaluate risks, and explore alternative solutions before contracts are awarded.

Calling repeatedly for a public–public partnership, Shiva elaborated that problems like water leakages, maintenance of pipelines, collection of revenue, and rationalization of tariffs could be handled without international intervention:

> We need to work towards a public–public partnership and not privatization. If the citizens are involved, they will by themselves ensure that there is no wastage of water and will keep a check on theft of water. The RWAs will

ensure that no resident steals water because it will cut into their entire supply. This is a much more cost effective and flexible option.

(Shiva Personal Interview 4 December 2004)

Through better management, including monitoring leaks and checking theft and corruption, about 50 percent of water problems would be solved, she noted. There are many alternative approaches to privatization from state run utilities in various different forms to cooperative schemes and community managed water systems. Other alternatives, Shiva suggested, include cleaning the Yamuna River, increasing ground water levels and reducing the level of wastage by involving the RWAs. Instead of increasing the cost for the common man or woman, it was suggested that to meet the huge deficit currently faced by the DJB, commercial users should be charged more.

Campaigners provided a plan whereby the RFSTE and the WWA could run the Sonia Vihar Plant on their own, saving the Government of Delhi crores of rupees in performance, management, and maintenance fees. Their research, published as *Delhi Jal Board Financial Sustainability is Possible through Public–Public Partnership*, was submitted to the Government on 8 September 2004 and proved that the DJB's financial sustainability could be made possible through public–public partnership and cooperative initiatives. Hence, using the huge potential of the new raised tariff structure to meet the DJB's requirements, India could avoid burdening citizens with enhanced loan liabilities from the World Bank's conditions. The only way to address the demand for an efficient, equitable, and cost effective water supply system, they asserted, was to make the public sector work more efficiently in a transparent and accountable manner. According to these campaigners, when the Government outsourced water treatment and distribution to the private sector, it shifted management of water supply to the management of contracts for water supply without addressing the institutional malaise. Building the Government's capacities to implement water policies remained extremely important but in the process of involving external actors it is largely left out of the international debate by those who prioritize water privatization.

The CFWD statement issued against tariff increases on 1 December 2004 sums up the position of the activists and the citizens:

The citizens of Delhi are committed to conservation and equitable use of our scarce, but precious water resources.

We are also committed to defend our fundamental right to water, which can only be protected through a public system, which treats water as a public good and essential service. Our right to water propels our pledge to keep Delhi's water supply in the public domain.

We condemn the anti-democratic, unjustified hike in tariffs announced by the Delhi government on November 30, 2004, which is a preparation for water privatization.

Citizens have offered models for public–private partnership to reduce waste and reduce costs, and provide safe, clean, affordable water for all.

The tariff increase carried out by the Delhi Jal Board on the World Bank's dictate will promote waste by the rich and put a burden on the poor. This "rationalization" might suit World Bank rationality to privatize and commodify the last drop of water. However, it goes against our culture and our constitutional rights.

The Government and World Bank are paving the way for Multinational Corporations (MNCs) like Suez to take control over our water. The contract for the Sonia Vihar plant has already been given to Suez-Degremont. With the tariff increase, a profitable "water market" is being created for MNCs.

As a brand new citizens' alliance of residents, environmental groups, religious groups, health groups, water workers, we will continue to work creatively and constructively to defend "water for life, not for profits." We will not allow our water to be hijacked. We will not let our democratic rights be bypassed. We will not let our fundamental right to water be eroded.

Toward that goal, we created the Citizens Front for Water Democracy on December 1, 2004, in response to the announcement of the tariff increase.

(RFSTE 2006)

This overview of the claims by which certain civil society actors engage in the water reform project demonstrates how they open up space to bring new voices and discourses into the policy process, but the extent to which they do so depends on the degree to which these spaces are used by or intersect with other spaces and actors. The empirical examples in this case study highlight the importance of looking away from policy statements and government documents that attribute particular roles to particular actors to try and understand the network and alliances that underpin the agencies of different actors to articulate alternative narratives of the water reform project and to act to change it. The success however, is determined by the ability to exert the power and influence to bring about a policy shift. Chapter 8 analyzes these dynamics in an attempt to understand these policy processes.

Notes

1 While water is not mentioned in the 1966 UN Covenant on Human Rights or in the 1948 Universal Declaration of Human Rights, it is inferred from Article 6 of the Covenant and Article 3 of the Universal Declaration, which grants every human beings the inherent right to life, and water is a basic natural life support for all.

2 The commons is generally defined as that which is not legally owned or controlled by private entities – things like air, public lands, oceans, and, yes, public water supplies. From another perspective, any public policy based on the principle of the commons seems foolish and anarchical in the eyes of many free enterprise proponents.

3 These include premier academic institutes like the Indian Institute of Technology (IIT) and the Indian Institute of Management (IIM), and activists like Vandana Shiva, Medha Patkar, and eminent economist H. M. Desarda.

4 The DJB has an enormous debt of Rs. 33 crores due to the delay in the commissioning of the plant and for the failure of DJB to supply raw water to the plant.

5 The current project of the Sonia Vihar Plant and the 24/7 project, they argue, will make water flow from the poor to the rich. This point was also raised among the DJB officials who felt that the DJB was giving preference to the elite and rich residing in South Delhi colonies over the poor and lower middle classes living in unauthorized colonies of Outer Delhi and rural areas of the capital. Even upper middle class residents have joined the fight against a process they call corrupt and unjust.

6 There is a traditional practice in India of putting of water huts (Piaos) in different parts of the country mostly in summer. Their purpose was to quench the thirst of a traveler who passes that way. These Piaos have long been part of Indian civilization and culture, and are established mostly by the affluent and rich or charitable institutions to provide for water in the summer heat when temperatures reach more than 100 degrees Fahrenheit.

8 Understanding the water policy process

This story of the making of a policy maps out the contemporary processes of water policy production in Delhi, which form an interesting and instructive illustration of reform within the overlapping of political spheres and the various pressures shaping water policy. In tracing the water policy process in the National Capital Territory (NCT) of Delhi, the case study revealed four major players, whose discourses played a major role in understanding these processes. These actors were: (1) the World Bank; (2) the Governments of India and the NCT of Delhi; (3) the consultants – PricewaterhouseCoopers (PwC); and (4) the Water Liberation Campaign (WLC) and the Citizens Front for Water Democracy (CFWD) that consisted of nongovernmental organizations (NGOs) and the affected communities.

Analysis of the competing discourses that emerged during the research investigation shows intense interaction and contestation in the interplay of power and resistance struggles, making the water policy process indeed complex and

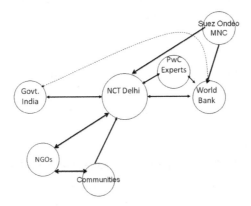

Figure 8.1 Actors in the Delhi water reform project.

dynamic. Figure 8.1 shows the complex interactions of these actors. In the discourses of power, an atmosphere of crisis was established through narratives of the irregularity, poor reliability, and scarcity of water supply that dominated the water sector in Delhi and thus necessitated water reforms in the first place. A network of experts – not only technicians but also administrators, strategists, and political advocates – was mobilized to drive this vision forward. These people became the core of a group that included both Indian and global experts. Change agents in the bureaucracy were appointed to important posts.

The global expertise came through the "invited participation" of the consultants, PwC, who were engaged to assess the status of the water sector and to provide inputs for remedial measures. These consultants were hired with the approval of the World Bank. The NGO Parivartan exposed the World Bank's intervention in the hiring process to which the World Bank responded that "the insinuation that the Bank attempted to favor PWC is completely unfounded – on the contrary, this is an excellent example of the Bank's close monitoring of the procurement process to ensure transparency and fair competition" (Carter 2005). Economic growth facilitated by these technical experts and the national and state Governments engaged in a donor–recipient relationship was seen to result in good governance and good infrastructure leading to poverty alleviation and growth.

The World Bank considers water as an infrastructural bottleneck that is an obstacle to development. Its narratives of water include bad management, bad quality that does not favor broad economic growth, and irregular and nonpotable water supply. On the positive side, its narrative stresses that continuous and good quality water supply based on the technological expertise and good management practices of private players would lead to development and poverty alleviation.

The Government of Delhi in this case echoed the rhetoric of the World Bank and the consultants, citing efficiency, quality of water, and 24/7 water supplies to the poor under the aegis of the science, technology, and skill available through the private players. This thinking echoes the rhetoric of transnational institutions that have entered the fray to provide the knowledge, technology, and skills necessary to manage the water sector. There is no doubt that there is an emerging market in water estimated to be worth over $2 billion (Shiva, *et al.* 2002: 11). This encouraged multinational corporations (MNCs) to move in under the Government's framework of "public–private partnership" (PPP). The 24/7 project and the Sonia Vihar Plant were initiated to address the problems of Delhi's water supply (e.g. intermittent and insufficient water supply, bad quality, availability, etc.) and aimed to improve the potability of the water, build the water supply infrastructure, and provide social and economic development as well as environmental sustainability. The water reform discourses incorporated these claims, and solutions were presented in the politically neutral terms of a technical approach.

This vision of the Government of Delhi was contested once information was leaked to the press about the Sonia Vihar Plant contract and the 24/7 project was exposed by Parivartan. The active engagement of the media, which latched onto the controversy and invested considerable coverage for some time, served to broaden the debate. The continuous press briefings from the Government and the NGO's comments made it a story that apparently had two sides. Various actors within the civil society pursued a variety of claims to challenge the policy process and negotiated their interests within the political spaces. For example, those who hailed from the religious and spiritual community staked claims of culture and religion over water; water users claimed tariff hikes affected their budgets; and scholars and academics called for the adoption of alternative local knowledge and technology. The developing debate on water reform resulted in a number of public reactions: symbolic protests, mass protests, and demonstrations, as well as media campaigning. Indian NGOs linked with international NGOs and the controversy also became part of the anti-globalization debate.

While the Government of Delhi claimed to take a participatory and people centered approach to the formulation of water policy, activists resisted and protested the changes, asserting that consultation at the policy formulation level has been merely pro forma and did not reflect the needs and aspirations of the people of Delhi.

As alternative claims and discourses have evolved, the water policy process has become increasingly complex, dynamic, and multiscalar, and this study has chronicled the competing discourses. Representatives of civil society lament the involvement of global policies in local practices while donor agencies feel that they have not dominated domestic policy making in India and that the Government of India's own recognition of the need to change is what led to the talk of a policy shift. Finally, the Government of Delhi claims that its policy shift was oriented towards benefitting of its citizens. What conclusions can be drawn from such a complex interweaving of actors and institutions, claims and counterclaims? The following section provides an overview analysis.

Dynamics of water policy processes

1   Water policy processes in the neoliberal era are an engagement in "deliberative exclusionary processes" on the one hand and "participatory processes" on

the other which represents a marriage of convenience between mainstream/ orthodox and participatory approaches. Hybridization of the narratives of power has led to the introduction of "participation" into reform strategies; the *Delhi Water Supply and Sewerage Sector Reform Project* report (DJB 2004) makes a special mention of it (ibid.: 8). Through the process of hybridization, mainstream narratives capture and selectively incorporate concepts generated by alternative development discourses. Yet orthodox approaches have continued to dominate the thinking of the Government of Delhi, perpetuating the assumption that technology and good management are the solutions to the water supply problems and that only private players like Suez, Saur, and Veolia/Vivendi – the water giants – can provide the expertise needed to unclog the bottlenecks in the infrastructure and build an efficient new system. Thus, as this study has shown, alternative discourses characterized by phrases like "bottom up," "people centered," and "participatory" have served merely to qualify but not alter foundational assumptions in any way. These terms remain very much in the realm of invited spaces and consultations in which the more powerful actors frame the way others are involved in the policy process. On the one hand, the process of incorporation into the power discourse can be regarded as a hegemonic strategy that defuses criticism by appropriating certain basic key phrases related to policy reform. On the other, it can also be cast as a successful incorporation of activists' perceptions that opens up possibilities for action and change. The way alternative discourses emphasize the strengths of "participatory" and "people centered" approaches requires further attention, consideration, and incorporation into the discourses of power.

2 The post-reform era has seen a major shift away from the discourses of equity and rights that characterized the approach to water in the pre-reform era. Previously water supply was a state led endeavor: the public sector supplied water to the city at subsidized rates because of the effect it had on the daily lives of people. In the post-reform era water began to be seen as part of the new economy and was constructed as the essential, defining trait necessary for a good life, sustainable development, and poverty alleviation. The National Water Policy (2002) and individual states' water policies, including Delhi's, frame water as a development problem allowing the solution – development through economic growth facilitated by capital investment in technological expertise in water infrastructure – to be presented as a self-evident truth defined in purely technical, politically neutral terms. The promotion of particular technical approaches lends further persuasiveness to a science based, expert driven policy discourse because framing it in a

technical approach leaves less room for opposition and hides the political intent of leaders and politicians. These changing constructions of water, framed by the business–science elite, have shaped Delhi's water reform policy in terms of an essentially economic and technological discourse. By promulgating stories that invite specific intervention, those in power have the ability to create frames of reference that define and bound what forms of knowledge count and whose versions, claims, and interests are legitimate. These often act as drivers of advocacy of particular policy instruments.

3 Policy processes are political by nature: a simplified overview of major shifts in the construction of water reforms and the absorption of some of the policy instruments are associated with different stages in the evolution of its discourses. Water reform has been constructed as a 24/7 water supply for the people of Delhi, facilitated by the involvement of the technologies and skills of MNCs and the inflow of global capital. The World Bank has opened up funding in water sector infrastructure development and paved the way for PPP. The process of reform, defined in terms of technical and managerial skills and projected as politically neutral, is billed as the only solution to water supply problems. The suggested reforms are structured by particular discourses that impose meanings that empower some and disempower others – an essentially political act. The political nature of policy in the case of Delhi has been camouflaged by the idioms of objectivity, neutrality, and rationality. The state projects itself as a competent regime that is dynamic, modern, and developmentalist and tries to shape policy making by raising concerns of regime legitimacy. While other actors, like the WLC and the CFWD, struggle to negotiate immediate policy outcomes in what is a clearly political process, the state attempts to define the terms of debate by presenting itself as a neutral arbiter, which it clearly is not. The rhetorical depoliticization of water reform policy by the mainstream actors shapes the way new actors are invited into the debate while at the same time obscuring the political nature of the policy and thus the range of options open to its opponents.

4 Transmission of knowledge through networks is an important aspect of policy making. Networks of power operated during the entire process of water policy reform in Delhi. Figure 8.2 demonstrates the linkages of power. An elite network of policy makers has come to dominate the post-reform era. As this study demonstrates, the Government of Delhi's vision echoes the rhetoric of the World Bank, the Government of India, PwC, and change agents within the bureaucracy. The continuity in the ideologies of this policy coalition is evident from the way water is ultimately framed in its narratives and taken up in its discourses. The Government of Delhi replicated global ideologies

in Delhi's local practices. The transmission of knowledge and the networks through which it occurs thus occupy an important place in this policy making process.

One can see in Figure 8.2 the complexities and dynamics within which the bureaucrats, laterals, politicians, consultants and the World Bank operate by being a part of the formal policy process in India in between operating as officials in global institutions where their training helps them transmit these ideologies into local practices. As the research demonstrates, different narratives about science, technology, and PPPs are associated with different networks of actors and emerge out of particular contexts and conditions. In the case of Delhi, these narratives have emerged from globalized connections, MNCs, and the political positioning of the Government of India and the Congress Party coalition in power. These core networks operate outside of the realm of democratic politics.

Science and technology are inseparable from the concept of modernization in the evolving orthodoxy of water. The knowledge required to make policy was constructed as a form of technical or scientific information, which would reveal the "truth" of the water situation. As the power of expertise and the transfer of technology came to underpin development policies, the knowledge of some groups of development actors (e.g. the PwC consultants and the World Bank) was defined as fundamentally more valid than the knowledge of others. As a result, alternative voices on the subject of development were rendered invisible and mute in pursuit of technically adequate solutions. Within the corridors of power the dissenting voices of the Bhartiya Janata Party (BJP) political party, the WLC, and the CFWD were neither heard nor incorporated into the Government of Delhi's policy at formulation level.

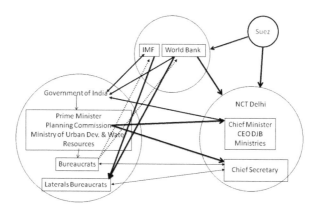

Figure 8.2 Networks of power.

5   While networks of power operated in incorporating global ideologies into local practices, there existed activist networks in the form of WLC and the CFWD. An interesting observation about these networks was that while they came from different sites of conflict along the River Ganga, they mobilized themselves as one voice in the Delhi water reform project (see Figure 8.3).

These multiple sites included groups from the Tehri Dam displaced population, the religious groups who resented the impounding of the sacred Ganga, people from the farmer communities who lost their land to the pipeline that transferred water to the Sonia Vihar Plant run by Ondeo Degremont, and the people of Delhi who would be victims of the 24/7 project. While their goals were initially different, they all mobilized under the WLC and the CFWD and challenged the state and corporations entering the scene. These networks helped in the formulation of an alternative discourse at the policy implementation level. The press was another actor that gave wide coverage to the controversy, specifically to the views of the activist networks.

Thus at the local level, in the implementation phase, networks of interests played an important role in shaping policy outcomes. From the Ganga liberators to the Resident Water Associations and women's organizations – some with more power; some with less power – a wide variety of actors interacted with each other in local arenas, contesting for realization of their claims against the claims of the state, and reflecting volumes of information and insights about power, complexity, and the political nature of the policy process.

6   An underlying theme in this analysis is the changing role of the state in the policy making process. In India, national policy debates have been influenced by, and in turn influence, narratives of global policy. Such narratives, which

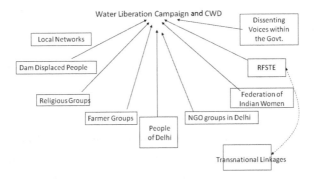

Figure 8.3 Networks of resistance.

often define both the assumed problem and the proposed solutions in neat, appealing storylines (cf. Roe 1991), are promulgated through key actors and their proposed networks. In discerning this complexity and tracking the dynamics of policy processes, Chapter 4 highlighted the personal, institutional, global, political, and business connections at work in the formulation of national water policy and revealed the interchange between local settings and global processes. A notable trend in the post-reform era has been the move toward state level (subnational) influence over policy making and the increased importance of the state level processes. Individual Indian states no longer dance to the tune of the center. They work and compete to attract foreign investment by international financial institutions that provide much support in infrastructure and incentives, and which are essential to keep the progressive and developmentalist visions of the politicians alive. Despite its neoliberal rhetoric, however, the Government of Delhi has not adopted a completely free market doctrine concerning its water policy reform. The state is intimately involved in supporting the enterprise. The partnership between the state and the private sector, with the backing of experts, is hailed as the new model for policy making, casting aside corrupt patronage politics of the "License Raj" and opening up a scientifically supported, economically sound set of polices by those who know best. But can such an elite technocratic vision of the state stand the test of actual implementation? As the case study demonstrates, the vision has stalled in some areas and gone ahead in others. This shows that while the interests of the new policy elite and their network are played out in the state, the possibilities for capturing the reins of power remain severely constrained. In part, volatile electoral politics and a vibrant civil society do set limits for the state. Diversion of resources to an MNC for management and distribution became a contentious electoral issue and entered the realm of mass politics. The state could not, in the final analysis, sell the 24/7 vision to the people/electorate, but it was able to move ahead with the Sonia Vihar Plant that enhanced supply of water to certain parts of Delhi.

7 Policy processes at the formulation level in India are state centric. This is evident both at the international level and at the state level. The role of the World Bank and its intervention in the policy process was documented in Chapter 5, but one cannot conclude that policy processes were totally dominated by the donor. Even though the Sonia Vihar and 24/7 projects exhibit global ideologies, the donors have not been the only ones to influence project decisions. Bureaucrats like Rakesh Mohan also shaped the policy, politicians like Rajiv Gandhi and Sheila Dixit promoted it; their ideas and interests

combined to influence the making of this policy. In fact even the donors expressed the opinion that the roots of domestic policy making in India are very strong and cannot be channeled by the donors alone. The Government of India, for example, had already signaled the shift in policy for Delhi by passing the loan for the Sonia Vihar project, so the external players were in effect leaning against an open door when they found their space within the process. The donors, however, do exercise some influence in the subnational dynamics in states like Delhi in the post-reform era because the states need donor investments to offset financial deficits and promote developmental projects. Various ideas, interests, and sources of power have shaped the dynamics of policy making in Delhi's water reform, and it remains an inter-action between domestic and external players.

The process dynamics reveal that few independent civil society organizations are engaged with the Government of Delhi and actively involved in influencing the policy process at the formulation level. Policy formulation in India is very much centralized and there is not much debate about the "policy on paper." Opposition is weak. It seems exceedingly important to remark that while water reform policy was under formation, it was within the control of actors such as PwC, tightlipped bureaucrats within the state government, and the World Bank. It was only when the blueprint of the reform was ready and the contract already awarded to Suez-Degremont that the policy was made known to agencies of civil society and other stakeholders.

The policy implementation stage marks the point at which the process has been contested, sabotaged, and manipulated in many different ways. It is in this space that opportunities for most policy spaces occurred. The resistance discourses and contestations in Delhi were built at this stage. The WLC and the CFWD, Parivartan, and other NGOs opened up policy spaces that enabled them to challenge pervasive orthodoxies, reframe the policy debate, and reconfigure relationships between actors. "Discourse coalitions" within policy arenas, like the dissenting voices in the state under the Water Workers Alliance (WWA) and other actors from outside like the WLC and the CFWD, have been instrumental in opening up these spaces – helping both to create new policy discourses and gain entry into policy arenas.

This appears to confirm that centralization limits broad participation in the policy making arena; the process does not become truly participatory until it reaches the implementation stage. The people are supposed to be partners in development with government; a centralized partnership requires that citizens have some say in the processes of agenda setting and local policy design. This is not happening. The partnerships that form in the

developmental stage tend to be between local Government and global institutions rather than with citizens. Despite enactment of the right to information,
the bureaucracy remains tightlipped in spelling out policy process details or
at best tries to remain anonymous, to allow any free and fair evaluation of
policy processes in India.

India has been a functional democracy for more than 50 years now: it has
a free press; a rich intellectual culture imbued with lively debate over issues
that affect society for good or ill; and a well-entrenched civil society engaged
at the local and global levels *inter alia* discussing, debating, and creating
spaces for activism. Such intense involvement by the press and the public
in the business of governing has carried considerable weight in addressing
basic issues in many areas of society, but the policy making process in water
reform has resisted such democratic inroads. The Indian water policy process has largely ignored social demands at the policy formulation level, and
new public policy suggestions have been limited, yet still the voices of the
citizenry have created space for dissent and, to some degree, have forced
changes in what began as an essentially top down process.

8   The policy process is actually an ongoing and iterative process, continually
subject to review and new initiatives, a point Sabatier (1987) makes with
his Advocacy Coalition Framework. He maintains that policy making is an
ongoing process with no clear demarcated beginnings or terminations. At the
simplest level, there is an ongoing interplay between policy development and
policy implementation. However, in the attempts to develop and implement
policy, new situations arise that demand reflection and adjustments, even
reformulation of policy. Within these iterations, there are also more complex articulations and feedback loops that further analysis should attempt to
capture. These complexities are especially evident in the interactions among
actors within the state and civil society, who represent a vast constellation
of competing interests and have taken a variety of positions on the issue of
water reform. Indeed, the sheer complexities of these interactions have forced
changes in the program at the *implementation* stage. The original intent of
the Delhi water reform policy was to augment the treated water supply to
accomplish the 24/7 vision. Currently, while the Sonia Vihar Plant has started
to function after three years of delay, the 24/7 project remains on hold. In
essence, the various actors have used their leverage to constrict to some
degree the space originally envisioned for global involvement. Although
the proposed reform project report still occupies a space on the Delhi Jal
Board's website and that of the World Bank in 2007, the Government of
Delhi has withdrawn its World Bank loan application. Reports proclaim that

the Government still maintains its sector vision on water but whether international players will be there or not is subject to review. While one would expect state actors to be able to preserve their policy objectives and condition the shaping of policy in practice, in the Delhi case, India's vibrant democracy was able to achieve a partial success in stalling an aspect of policy change that was going to affect the daily lives of 14 million people in Delhi. What do these events mean for policy analysis?

Two approaches to policy analysis

Since the growth of the process oriented branch of policy studies is a relatively recent phenomenon, one that has taken place over the last 15 years, the literature is dominated by examples from researchers from the United States and to some extent Great Britain rather than from developing countries (Mooij 2003). The study of policy process literature is dominated by two approaches to policy analysis: (1) the linear or state centered model and (2) the horizontal or society centered model.

State centered accounts of policy making follow a linear model of policy analysis with the state as the key actor. They focus on bureaucratic politics, epistemic communities, professional repertoires, and management models. According to this approach, policy constitutes the decisions made by those with responsibility for a given policy area, and these decisions usually take the form of statements or formal positions on an issue, which are then executed by the bureaucracy.

The linear model, which was dominant during 1970s and 1980s (Deleon 1999: 23), presumes that policy process moves through agenda setting, policy formulation, implementation, and evaluation. Decision making is seen as the key moment in the policy process. Once a policy decision is made, the entire bureaucratic machinery starts operating simply for its execution or implementation for its

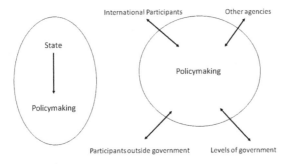

Figure 8.4 Linear and horizontal models of policy making.

success. In case of failure, the blame is attributed to "bottlenecks," interference, or "lack of will," under the assumption that external factors have no impact on the "policy proper," i.e. the point of decision making (see Chapter 2).

The horizontal or society centered accounts of policy making shift the focus away from "policy makers" to a much broader constellation of actors who engage in various ways with the process of making and shaping policy. This approach, its proponents believe, brings the dynamics of policy processes into clearer view. Through the use of analytical tools, such as models of actor networks, interface analysis, class analysis, bargaining models, and discursive coalitions, the society centered approach seeks to highlight the relational dimension of policy making, focusing on processes of negotiation and contestation, and on networks, alliances, and coalitions through which policies are shaped – from "policy networks" to "policy communities" (Haas 1992; Smith 1993a). This model assumes that policy is made through the interaction of competing interest groups, and that each policy decision is a separate event unrelated to other policy decisions. By exploring the strategies and tactics that particular kinds of actors draw on to attempt to shape the direction of policy, accounts of this kind can offer insights that undermine the linear policy narrative and the causal assumptions that are attached to specific avenues of implementation only after a new policy agenda has been set.

A society centered discursive analysis focuses on the way in which issues are framed. Shifting narratives of the causes of and solutions to issues both produce and drive policy processes, making available and circumscribing spaces in which – in the present study – different versions of water policy reform knowledge can be articulated and mobilized. First, Apthorpe argues, "Rival ways of naming and framing set policy agendas differently" (1996: 24). Highlighting the style, form, and language used in the construction of policy statements and in the interactions that shape policy processes, strategies such as deconstruction and narrative analysis prove valuable in policy analysis. Second, this approach to the analysis of policy discourses has a wider purview. An analysis of framing extends from semiotic or narrative analysis of policies themselves to the ways in which policy as discourse frames the roles of different actors in the policy process. The analysis of discourses and framing devices used in policy deliberations highlights the ways power, knowledge, and political spaces unfold, operate and become embedded in the framing of policy.

Concluding observations on water policy processes in the post reform era

Adopting the society centered approach, one must analyze the policy process as multiple spaces of contestation involving complex configurations of actors, discourses, and knowledge in order to understand the ways in which, in this study, water knowledge affects water policies. What is apparent in the contextual "messiness" of policy production, however, is that studies of water governance and policy production cannot remain solely focused on a global–national duality, nor can policy analysis models remain fixated on the state, centered on a top down approach or on a society centered, bottom up approach. Both state centered and society centered accounts of policy development and implementation are needed to provide a full picture of the complexities involved in the process.

Figure 8.5 analyzes these interactions in the policy process of water sector reform in Delhi. At the state level are generally the donors, acting both to strengthen the state's capacity for effective policy implementation and also to provide the enabling environment for decentralization of policy activity. But on the level of civil society there exist other actors with specific projects and interests that involve contestation and power plays over the control and use of water resources in Delhi.

Policy making involves both the linear and the horizontal dimensions – not only power, knowledge and action relationships in the horizontal construction of policy spaces but also the vertical construction of power relations that are reinforced or recreated through interactions at different levels of the policy process. In the case of Delhi one can observe the secrecy with which the Governments of India and Delhi worked, especially in the award of contract to Ondeo Degremont.

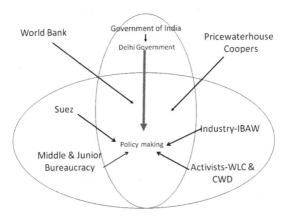

Figure 8.5 Policy production in the post-reform era.

There also exists a clear transmission of knowledge from the national to subnational and local levels of the paradigm shifts in water policy. In the above analysis of the nature of policy production, in addition to the complexities mentioned heretofore, there are pressures emanating from a variety of local, subnational, national, and global scales, expressed by a variety of voices, and existing within varying dynamics of inclusion and exclusion. The voices of resistance were included in the decision making process involving the 24/7 project but excluded from that of the Sonia Vihar project. In matters of water policy process, while there are pressures from "above" – global pressures offering universal solutions – there are simultaneously pressures from "below" among civil society and other water stakeholders in the sociocultural realities of the Indian state who contest, sabotage, and manipulate implementation in many different ways. It is here that policy spaces are most likely to open. As these competing narratives of power and resistance evolve, water reform policy itself evolves in the interplay of complex and dynamic multiscalar processes. In sum, policy making reflects continuity in centralization processes at the level of formulation while most of the contests take place at the implementation level. There is an interaction of the linear and horizontal models in policy making. At the formulation level, the model seems to be one of a top down approach within which ideas, interests and power interact while at the implementation level it leans more towards contestation and engagement.

Thus, policy is complex, and dynamic. It involves a complex configuration of interests among a range of differently positioned actors, whose individual agency matters but whose interactions are shaped by power relations. By highlighting how polices are being made in practice, this study has revealed the importance of the relationships between global, national, and subnational policy processes, embedded as they are in complex networks of divergent and overlapping interests that create a particular political economy of knowledge and power. By tracing the origin of the water policy in the state of Delhi and examining its historical, economic, and political context and the key debates in relation to Delhi's engagement with global pressures, as well as pressures emanating from local, subnational, and national imperatives, one can define what policy making means in practice. And, in a larger sense, one can answer the questions: What is policy? What are the processes – technical, political, intellectual, and social in policy making? What do these inputs suggest about the nature of policy making in so important sector as water – seen as the key to economic growth and prosperity in the arena of development and central to the new economy era?

Policy processes are much more complex than previously believed when seen in response to the changing contexts of the new economy era of liberalization, privatization, and deregulation. The new politics of policy making is far removed from the previous understandings of the policy process in India, dominated since the post-independence period by centralized planning, where states danced to the tune of the center and the private sector was not a major player. Water policy in particular was characterized by centralization and hierarchy, rooted in the post-independence planned development approach with a water resources bureaucracy dominated by civil engineers (Mollinga 2005). A new model for policy making has emerged, quite apart from the "License Raj," with the backing of the policy elite, experts from science and technology, global institutions like the World Bank, and the private sector. Policy space has emerged in the form of events and actors like the WLC and the CFWD challenging pervasive orthodoxies to reframe the debate and reconfigure relationships between actors. This process provides an important entry point to understanding the dynamics of alternative narratives and agendas shaping urban water reform policy. Policies are shaped by these competing narratives, informed by divergent interests, and articulated by different discourse coalitions based on power and their claims of knowledge. Existing concurrently as a rational technical solution as well as a political instrument, it is within this complex overlap of competing interests, imperatives, and pressures, rather than under the aegis of the state alone, that water policies are produced, contested, implemented, and reformed.

References

Agarwal, A., Narain, S. and Khurana, I. (2001) *Making Water Everybody's Business: Practice and Policy of Water Harvesting*. New Delhi: Centre for Science and Environment.

Agarwal, P., Gokarn, S. V., Mishra, V. K., Parikh, S. and Sen, K. (1995) India: Crisis and response. In A. Pradeep (ed.) *Economic Restructuring in East Asia and India*. New York: St Martin's Press, pp. 159–66.

Ahluwalia, M. S. (1998) Towards Liberalisation and Globalisation. In V. Ramachnandran (ed.) *Rajiv Gandhi's India: A golden jubilee retrospective, Vol II, Economics, Proceedings of the 2nd symposium. November 21–4, 1994*. New Delhi: UBS Publishers, p. 203.

Ahluwalia, M. S. (1997) Financing private infrastructure: Lessons from India. In H. Kohli, A. Mody and M. Walton (eds.) *Choices for Efficient Private Provision of Infrastructure in East Asia*. Washington, DC: World Bank.

Ahluwalia, M. S. (1994) Commentary on India's reform. *Journal of Business* 29 (1): 17.

Ali-Bogaert, M. A. (1997) Imagining alternatives to development: A case study of the Narmada Bachao Andolan in India. Unpublished Ph.D. dissertation, Kent State University.

Allchin B. (1998) Early man and environment in South Asia: 10,000 BC–500 AD. In R. Grove, *et al.* (eds.) *Nature of the Orient: The Environmental History of South and Southeast Asia*. Delhi: Oxford University Press, pp. 29–50.

Anderson, J. E., Brady, D. W., Bullock, C. S. III and Stewart, J. S. Jr. (1984) *Public Policy and Politics in America*. Montgomery, CA: Brooks-Cole.

Atkinson, M. M. and Coleman, W. D. (1992) Policy, communities, policy networks and the problems of governance. *Governance* 5: 154–80.

Arora, D. (2002) Public policy analysis: Addressing the contextual challenges. *Indian Social Science Review* 4 (1): 45–68.

Arora, D. (1993) On the tragedy of public domain: Corruption, victimization and the new policy regime. *The Indian Journal of Public Administration* 39 (1): 383–94.

Arora, G. K. (1998) Towards Liberalisation and Globalisation. In V. Ramachnandran (ed.) *Rajiv Gandhi's India: A golden jubilee retrospective, Vol II, Economics, Proceedings of the 2nd symposium. November 21–4, 1994*. New Delhi: UBS Publishers, p. 197, 200.

ADB – Asian Development Bank (2001) *Water For All: The Water Policy of the Asian Development Bank*. Available online at: http://www.adb.org/documents/policies/water/ (accessed 21 November 2004).

Babbie, E. (1998) *The Practice of Social Research*. Belmont: Wadsworth.

Barker, C. (1996) *The Health Care Policy Process*. London: Sage.

Barve, V. (2002) Pani puravthyachya khajgikaranacha dav asa udhala [This is how we busted the water privatization]. *Andolan* September: 6–10.

Basham A. L. (1967) *The Wonder That Was India.* New York: Grove Press.

Benford, R. D. and Snow, D. A. (2000) Framing processes and social movements: An overview and assessment. *Annual Review of Sociology* 26: 11–39.

Bhaduri, A. and Nayyar, D. (1996) *An Intelligent Person's Guide to Economic Liberalization.* New Delhi: Penguin Books.

Bhattacharya, S. (1975) India's first private irrigation company. *Social Scientist* 4 (3) October: 35–55.

Blaike, P. and Soussan, J. G. (2001) Understanding policy processes. Working Paper. Leeds: University of Leeds.

Braybrooke, D. and Lindblom, C. (1963) *A Strategy of Decision.* New York: Free Press.

Briscoe, J. (2005) *India's Water Economy: Bracing for a Turbulent Future.* Washington, DC: World Bank.

Briscoe, J. (1997) Managing water as an economic good: Rules for reformers. Keynote Paper to the International Committee on Irrigation and Drainage Conference on Water as an Economic Good, Oxford, September. Available online at: http://rru.worldbank. org/Documents/PapersLinks/994.pdf (accessed 27 February 2009).

Callon, M. (1986) The sociology of an actor-network: The case of the electric vehicle. In M. Callon, J. Law and A. Rip Houndmills (eds.) *Mapping the Dynamics of Science and Technology: Sociology of Science in the Real World.* London: Macmillan.

Canepa, C. (2004) *Government Action, Community Organizing and Technology: Improving the Delivery of Water and Sewerage Services in Delhi.* Research Report for the Program on Human Rights and Justice. Cambridge, MA: MIT Press.

Carter, M. (2005) The World Bank's role in the Delhi Water Supply and Sewage Project. 29 July. Available online at: http://web.worldbank.org/WBSITE/EXTERNAL/ COUNTRIES/SOUTHASIAEXT/0,contentMDK%3A20600280~menuPK%3A15884 3~pagePK%3A146736~piPK%3A146830~theSitePK%3A223547,00.html (accessed 11 July 2006).

Cerny, P. G. (1990) *The Changing Architecture of Politics: Structure, Agency and the Future of the State.* London: Sage.

Chakravarti, R. (1998) The creation and expansion of settlements and management of hydraulic resources in ancient India. In R. Grove, *et al.* (eds.) *Nature of the Orient: The Environmental History of South and Southeast Asia.* Delhi: Oxford University Press.

Clay, E. and Schaffer, B. (eds.) (1984) *Room for Maneuver: An Exploration of Public Policy in Agriculture and Rural Development.* London: Heinemann.

Cobb, R. and Elder, C. (1972) *Participation in American politics: The Dynamics of Agenda-Building.* Baltimore, MD: Johns Hopkins University Press.

Colebatch, H. K. (1998) *Policy.* Minneapolis: University of Minnesota Press.

Collins, A. (2000) *Poverty Reduction Strategies. What Have We Learned So Far?* Brussels: Eurodad.

Congress Party/Indian National Congress. *Election Manifesto: General Election 1991.* New Delhi: AICC.

Considine, M. (1994) *Public Policy: A Critical Approach.* Melbourne: Macmillan.

CII – Confederation of Indian Industry (2005) Water Summit. Press Release. New Delhi: CII.

CII (1991) *Annual Report.* New Delhi: CII.

Corbridge, S. and Harriss, J. (2000) *Reinventing Indian Liberalization, Hindu Nationalism and Popular Democracy.* Cambridge: Polity Press.

Crow, B. with Lindquist., A and Wilson, D. (1995) *Sharing the Ganges: The Politics and Technology of River Development*. New Delhi: Sage.

Currie, B. (2000) *The Politics of Hunger in India: A Study of Democracy, Governance and Kalahandi's Poverty*. Basingstoke: MacMillan.

CGWB – Central Ground Water Board (1999) *Status of Ground Water Quality and Pollution Aspects in NCT-Delhi*. New Delhi: CGWB.

CPCB – Central Pollution Control Board (2004) Aquazur V.F. Available online at http://www.cpcb.nic.in/oldwebsite/AR2004/ar2004-ch7.htm (assessed 7 November 2006).

CSE – Centre for Science and Environment (2003) *Water in Delhi: Rain Water Harvesting*. New Delhi: CSE.

Daga, S. *Private Supply of Water in Delhi*. CCS Working Paper No. 0059. Available online at http://www.ccsindia.org/ccsindia/policy/trans/studies/wp0059.pdf (accessed 21 October 2006).

Dash, K. (1999) India's international monetary fund loans: Finessing win-set negotiations within domestic and international politics. *Asian Survey* 39 (2): 884–907.

David, S. (2005) Whose water is it anyway? *India Today*. Living Media India Ltd.

Dearing, J. W. and Rogers, E. M. (1996) *Agenda Setting*. Thousand Oaks, CA: Sage.

DeLeon, P. (1999) The stages approach to policy process: What has it done? Where is it going? In P. A. Sabatier (ed.) *Theories of the Policy Process*. Colorado: Westview Press, pp. 19–23.

Delhi Statistics (2007) Available online at http://www.indianbusiness.nic.in/know-india/sattes/delhi.htm (accessed 13 August 2007).

Deveraj, R. (2002) *Bypassing Community Rights: The National Water Policy*. Available online at: http://infochangeindia.org/200201015973/Water-Resources/Books-Reports/Bypassing-community-rights-The-National-Water-Policy.html

Dharmadhikary, S. (2003) Privatisation not the answer. *The Hindu*, 28 January.

Dixit, S. (2005) Delhi water not to be privatized. Interview in newswire. Indo-Asian New Service, 23 August.

Dixit, S. (2005) Water supply outsourcing to be an open book: CM. Interview. Delhi Newsline, 17 August. Available online at: http://cities.expressindia.com/fullstory.php?newsid+144242

DJB – Delhi Jal Board (2004) *Delhi Water Supply and Sewerage Sector Reform Project*. New Dehli: DJB. Available online at: http://www.delhijalboard.nic.in/djbdocs/reform_project/docs/docs/doc_project_prep_docs/introduction/DJB-ReformProject%20-%20Final.htm (accessed 15 September 2005).

Dobuzinskis, L. (1992) Modernist and postmodernist metaphors of the policy process: Control and stability vs. chaos and reflexive understanding. *Policy Sciences* 25: 355–80.

Dowding, K. (1995) Model or metaphor? A critical review of the policy network approach. *Political Studies* 43: 136–58.

Dreyfus, H. L. and Rainbow, P. (1982) *Michael Foucault:Beyond Structuralism and Hermeneutics*. Brighton, UK: Harvester.

Dror, Y. (1964) Muddling through: Science or inertia? *Public Administration Review* 24: 153–7.

Dryzek, J. (1997) *The Politics of the Earth: Environmental Discourses*. Oxford: Oxford University Press.

DUEIIP – Delhi Urban and Environmental Infrastructure Improvement Project (2001) *Delhi 21*. New Dehli: Government of India, Ministry of Environment and Forest and Government of the National Capital Territory of Delhi, Planning Department.

Available online at http://delhiplanning.nic.in/Reports/Delhi21/Delhi-21.pdf (accessed 7 November 2006).

Easton, D. (1965) *A Framework for Political Analysis*. New Jersey: Prentice Hall.

Echeverri-Gent, J. (1993) *The State and the Poor: Public Policy and Political Development in India and the United States*. Berkeley, CA: University of California Press.

Edwardes, M. (1967) *British India, 1772–1947*. London: Sidgwick and Jackson.

EGCIP – Expert Group on the Commercialization of Infrastructure Projects (1996) *The India Infrastructure Report: Policy Imperatives for Growth and Welfare*. New Delhi: Government of India.

Elhance, A. P. (1999) *Hydro Politics in the Third World*. Washington, DC: United States Institute of Peace Press.

Emerson, R., Fretz, R. and Shaw, L. (1995) *Writing Ethnographic Field Notes*. Chicago: University of Chicago Press.

Etzioni, A. (1967) Mixed-scanning: A "third" approach to decision-making. *Public Administration Review* 27: 385–92.

Finance Minister Budget Speech (2005–6) Available online at http://delhiplanning.nic.in/ Budget %20Speech/2005-06/Budget%20Speech%202003–04.pdf (accessed 7 March 2005).

Finger, M. and Allouche, J. (2002) *Water Privatization*. New York: Spon Press.

Fischer, F. (2003) *Reframing Public Policy: Discursive Politics and Deliberative Practices*. New York: Oxford University Press.

Fischer, F. and Forestor, J. (1993) *The Argumentative Turn in Policy Analysis and Planning*. London: University College London Press.

Foucault, M. (1980) *Power/Knowledge: Selected Interviews and Other Writings 1972–77*. New York: Pantheon.

Foucault, M. (1979) *The History of Sexuality, Part One: An Introduction*. London: Allen Lane.

Foucault, M. (1977) *Discipline and Punishment*. London: Allen Lane.

Frankel, F. (1978) *India's Political Economy of Development*. Princeton: Princeton University Press.

Gandhi, R. (1988) Inaugural Address to the All India Businessmen's Convention: New Delhi. In R. Gandhi, *Selected Speeches and Writings II*. New Delhi: Government of India, p. 65.

Gandhi, R. (1986) *Selected Speeches and Writings*. New Delhi: Government of India.

Gasper, D. and Apthorpe, R. (1996) Introduction: Discourse analysis and policy discourse. *European Journal of Development Research* 8 (1): 1–15.

Gerth, H. and Mills, C. W. (1991) Introduction. In H. Gerth and C. W. Mills (trans. and eds.) *From Max Weber: Essays in Sociology*. London: Routledge, pp. 3–74.

Giddens, A. (1990) *The Consequences of Modernity*. Cambridge: Polity Press.

Gilmartin, D. (1994) Scientific empire and imperial science: colonialism and irrigation technology in the Indus basin. *Journal of Asian Studies* 53 (4), November: 1127–49.

Goldman, M. (2005) *Imperial Nature: The World Bank and Struggles for Social Justice in the Age of Globalization*. New Haven, CT and London: Yale University Press.

Gordon, I., Lewis, J. and Young, J. (1993) Perspectives on policy analysis. In M. Hill (ed.) *The Policy Process*. Hemel Hemstead, UK: Harvester Wheat Sheaf.

Grillo, R. (1997) Discourses of development: The view from anthropology. In R. Stirrat and R. Grillo (eds.) *Discourses of Development: Anthropological Perspectives*. Oxford: Berg.

Grindle, M. and Thomas, J. (1991) *Public Choices and Policy Change: The Political*

Economy of Reform in Developing Countries. Baltimore: Johns Hopkins University Press.

Gulati, A., Meinzen-Dick, R. and Raju, K. V. (1999) *From Top Down to Bottom Up: Institutional Reforms in Indian Canal Irrigation.* Delhi: Institute of Economic Growth.

Gyawali, Dipak (2000) Nepal-India water resource relations. In I. Zartman and J. Rubin (eds.) *Power and Negotiation.* Michigan: University of Michigan, pp. 129–54.

Gyawali, D. and Dixit, A. (1999) Mahakali impasse and Indo-Nepal water conflict. *Economic and Political Weekly* February 27: 553–64.

Haas, P. (1992) Introduction: Epistemic communities and international policy formulation. *International Organization* 46: 1–36.

Habermas, J. (1980) *Legitmation Crisis.* London: Heinemann.

Haggard, S. and Kaufman, R. R. (eds.) (1992) *The Politics of Economic Adjustment: International Constraints, Distributive Conflict and the State.* Princeton: Princeton University Press.

Haggard, S. and Webb S. B. (1993) What do we know about the political economy of economic policy reform? *The World Bank Research Observer* 8 (2): 143–68.

Hajer, M. (1995) *The Politics of Environmental Discourse: Ecological Modernization and Policy Process.* Oxford: Clarendon.

Harriss, B. (1988) Policy is what it does: State trading in rural South India. *Indian Journal of Public Administration and Development* 8 (2): 151–60.

Harriss, J. (1985) The state in retreat? Why has India experienced such half-hearted liberalisation in the 1980s? *IDS Bulletin* 18 (4): 31–8.

Held, D. (1995) *Democracy and the Global Order:From Modern State to Cosmopolitan Governance.* Stanford: Stanford University Press.

Hindustan Times (2004) But Dikshit calls it a dream come true, 30 September.

Hjern, B. and Porter. D. (1981) Implementation structures: A new unit of administrative analysis. *Organization Studies* 2: 211–17.

Hogwood, B. and Gunn, L. (1984) *Policy Analysis for the Real World.* Oxford: Oxford University Press.

Horowitz, D. (1989) Is there a third world policy process? *Policy Sciences* 22: 197–212.

Houtzanger, P. (1999) Collective action and patterns of political authority: rural workers, church and the state in Brazil. IDS Draft Paper. Sussex, UK: IDS.

Howlett, M and Ramesh, M (1998) Policy subsystem configurations and policy changes: Operationalising the post positivist analysis of the politics of the policy process. *Policy Studies Journal* 26 (30): 461–81.

Howlett, M. and Ramesh, M. (1995) *Studying Public Policy.* Toronto: Oxford University Press.

Imperial Gazetteer of India (1908) Vol. III. Oxford: Clarendon Press.

Iyer, R. R. (2008) *Toward Water Wisdom: Limits, Justice and Harmony.* New Delhi: Sage.

Iyer, R. R. (2003) *Water: Perspectives, Issues, Concerns.* New Delhi: Sage.

Iyer, R. R. (2000) *Charting a Course for the Future.* New Delhi: Centre for Policy Research.

Jenkins, R. (1999) *Democratic Politics and Economic Reform in India.* Cambridge: Cambridge University Press.

Jenkins, W. (1978) *Policy Analysis: A Political and Organizational Perspective.* London: Martin Robertson.

Jessop, B. (1999) The changing governance of welfare. *Social Policy and Administration* 33 (4): 348–59.

Jha D. N. (1998) *Ancient India in Historical Outline.* Manohar: New Delhi.

Jordan, G. (1990) Sub-governments, policy communities and networks: Refilling the old bottle? *Journal of Theoretical Politics* 2: 319–18.

Juma, G. and Clarke, N. (1995) Policy research in Sub-Saharan Africa. *Public Administration and Development* 15: 121–37.

Kahler, M. (1990) Orthodoxy and its alternatives: Explaining approaches to stabilization and adjustment. In J. Nelson (ed.) *Economic Crisis and Policy Choice.* Princeton: Princeton University Press, Ch. 2.

Kanyinga, K. (1998) Contestation over political Spaces: The state and the demobilization of party politics in Kenya. Working Paper 98/12. Copenhagen: Center for Development Research.

Kapur, S. L. (2005) 50 years of Indian independence. *The Tribune Special Supplement.* Available online at http://www.tribuneindia.com/50yrs/kapur.htm (accessed 9 November 2007).

Kaur, N. (2003) Privatizing water. *Frontline* 20(18), 30 August–12 September. Kaviraj, S. (1997) The modern state in India. In M. Doornbos and S. Kaviraj (eds.) *Dynamics of State Formation. India and Europe Compared.* Indo-Dutch Studies on Development Alternatives 19. New Delhi: Sage, pp. 225–50.

Keeley, J. and Scoones, I. (2003) *Understanding Environmental Policy Processes: Cases from Africa.* London: Earthscan.

Keeley, J. and I. Scoones, I (2000) Environmental policy-making in Zimbabwe: Discourses, science and politics. IDS Discussion Paper. Sussex, UK: IDS.

Keeley, J. and Scoones, I. (1999) Understanding environmental policy processes: A review. IDS Working Paper No. 89. Sussex, UK: IDS. Available online at http://www.ids.ac.uk/ids/publicat/wp/wp89.pdf. (accessed 23 May 2005).

Kejriwal, A. (2005) Protest against water privatization. *The Hindu,* 12 July.

Khagram, S. (2004) *Dams and Development: Transnational Struggles for Water and Power.* Ithaca: Cornell University Press.

Kingdon, J. (1984) *Agendas, Alternatives and Public Choices.* Boston: Little, Brown.

Kling, E. H. (1996) Analyzing and managing policy processes in complex networks: A theoretical examination of the concept policy networks and its problems. *Administration and Society* 28 (1): 90–119.

Kochanek, S. A. 1996. Liberalization and business lobbying in India. *Journal of Commonwealth and Comparative Politics* 34 (3): 155–73.

Kohli, A. (2001) *The Success of India's Democracy.* Cambridge: Cambridge University Press.

Kohli, A. (1991) *Democracy and Discontent: India's Growing Crisis of Governability.* Cambridge: Cambridge University Press.

Kohli, A. (1989) Politics of economic liberalization in India. *World Development* 17 (3): 305–28.

Kosambi, D. D. (1965) *The Culture and Civilization of Ancient India in Historical Outline.* London: Routledge and Kegan Paul.

Lal, S. (2005) There's a hole in the bucket. *Times of India,* 26 October.

Lasswell, H. D and Kaplan, A. (1970) *Power and Society.* New Haven: Yale University Press.

Latour, B. (1987) *Science in Action: How to Follow Scientists and Engineers Through Society.* Cambridge, MA: Harvard University Press.

Lele, S. and Menon, A. (2005) Draft NEP 2004: A flawed vision. *Seminar* 547, March: 55–62.

Levitt, T. (1985) The globalization of markets. In A. M. Kantrow (ed.) *Sunrise … Sunset: Challenging the Myth of Industrial Obsolescence.* New York: John Wiley, pp. 58–68.

Li, T. (2002) Government through community in an age of neoliberalism. Paper. Anthropology Department, Santa Cruz: University of California.

Lindblom, C. (1959) The science of muddling through. *Public Administration Review* 19: 79–84.

Lindblom, C. E. (1979) Still muddling, not yet through. *Public Administration Review* 39: 517–26.

Lipsky, M. (1980) *Street Level Bureaucrat: Dilemmas of the Individual in Public Services.* New York: Russel Sage Foundation.

Llorente M. and Zerah, M. H. (2003) Urban water sector: Formal versus informal suppliers in India. *Urban India* 22 (1), January–June: 1–15.

Long, N. and Long, A. (eds.) (1992) *Battlefields of Knowledge: The Interlocking of Theory and Practice in Social Research and Development.* London: Routledge.

Maloney, C. and Raju, K. V. (1994) *Managing Irrigation Together: Practice and Policy in India.* New Delhi: Sage.

Manor, J. (1995) The political sustainability of economic liberalization in India. In R. Cassen and V. Joshi (eds.) *India: The Future of Economic Reforms.* New Delhi: Oxford University Press, pp. 341–61.

Manor, J. (1993) *Power, Poverty and Poison: Disaster and Response in an Indian City.* New Delhi: Sage Publications.

Manor, J. (1987) Tried, then abandoned: Economic liberalisation in India. *IDS Bulletin* 18 (4): 39–44.

Marcus, G. E. (1995) Ethnography in/of the world system: The emergence of multi-sited ethnography. *Annual Review of Anthropology* 24: 95–117.

Mathur, K. (2001) Governance and alternative sources of policy advice: The case of India. In K. Weaver and P. Stares (eds.) *Guidance for Governance: Comparing Alternative Sources of Public Advice.* Tokyo: Centre for International Exchange and Brookings Institute.

Mathur, K. and Jayal, N. J. (1993) *Drought, Policy and Politics in India: The Need for a Long Term Perspective.* New Delhi: Sage.

McCully, P. (2001) *Silenced Rivers: The Ecology and Politics of Large Dams.* London: Zed Books.

McGrew, A. and Lewis, P. (eds.) (1992) *Globalization and the Nation State.* Cambridge: Polity Press.

Mehdudia, S. (2005a) Water sector reforms a recipe for disaster. *The Hindu*, 3 November.

Mehdudia, S. (2005b) Aruna Roy turns down World Bank invitation. *The Hindu*, 25 November.

Mehdudia, S. (2000) Sheila Dixit plays slander campaign. *The Hindu*, 29 November.

Ministry of Water Resources. Water in the Indian Constitution. Available online at: http://www.wrmin.nic.in/index2.asp?sublinkid=404&langid=1&slid=299 (accessed 30 August 2005).

Mohan, R. (2005) Letter to Dr V. Shiva, Convenor, Citizens for Water Democracy and Director, RFSTE. In RFSTE, *Privatization of Water.* Research Report. New Delhi: RFSTE.

Mohanty, N. (2005) Moving to scale. Background Paper for *India's Water Economy.* Washington, DC: World Bank.

Mollinga, P. P. (2005) The water resources policy process in India: Centralization, polarisation and new demands on governance. Working Paper No. 7. Bonn: Center for Development Research, Department of Political and Cultural Change, University of Bonn.

Mollinga, P. P. and Bolding, A. (eds.) (2004) *The Politics of Irrigation Reform: Contested Policy Formulation and Implementation in Asia, Africa and Latin America.* Global Environmental Governance Series. Surrey, UK: Ashgate.

Mollinga, P. P., Doraiswamy, R. and Engbersen, K. (2001) The implementation of participatory irrigation management in Andhra Pradesh. *International Journal on Water* 1 (3/4): 360–79.

Mooij, J. (2003) Smart governance? Politics in the policy process in Andhra Pradesh India. Working Paper No. 228. London: Overseas Development Institute, October.

Mooij, J. (1999a) *Food Policy and the Indian State: The Public Distribution System in South India.* New Delhi: Oxford University Press.

Mooij, J. (1999b) Food policy in India: The importance of electoral politics in policy implementation. *Journal of International Development* 2: 625–36.

Mooij, J. and Vos, D. (2003) Policy processes: An annotated bibliography on policy processes, with particular emphasis on India. Working Paper No. 221. London: Overseas Development Institute, July.

Mukherjee, P. (1998) Towards Liberalisation and Globalisation. In V. Ramachnandran (ed.) *Rajiv Gandhi's India: A golden jubilee retrospective, Vol II, Economics, Proceedings of the 2nd symposium. November 21–4, 1994.* New Delhi: UBS Publishers, p. 207.

Narsalay, R. (2003) Who controls rights over water in an unfolding GATS regime? A case study of India. Paper prepared for the Research Project on Linking the WTO to the Poverty Reduction Agenda, June. Available online at: http://www.gapresearch.org/governance/NarsalayGATSandwaterpaperjune03.pdf (accessed 8 July 2006).

National Water Policy (2002) New Delhi: Ministry of Water Resources, Government of India. Available online at http://cwc.nic.in/nwp2002.pdf NWP 2002 (11 October 2005).

National Water Policy (1987) New Dehli: Ministry of Water Resources, Government of India. Available online at http://www.cgwb.gov.in/documents/nwp_1987.pdf (11 October 2005).

Nayar, B. R. (2001) *Globalization and Nationalism: The Changing Balance in India's Economic Policy, 1950–2000.* New Delhi: Sage.

News Track (2006) TV Broadcast. New Delhi: Living Media Ltd., 29 May.

Nigam, A. (2001) Dislocating Delhi: A city in the 1990s. Available online at http://www.sarai.net/publications/readers/01-the-public-domain/040–6industry.pdf (accessed 21 October 2006).

Nustad, K. and Sending, O. (2000) The instrumentalization of development knowledge. In D. Stone (ed.) *Banking on Knowledge: The Genesis of the Global Development Network.* London: Routledge, pp. 44–62.

Offe, C. (1984) *Contradictions of the Welfare State.* London and Sydney: Hutchinson.

Ohmae, K. (1995) *The End of the Nation State: the rise of regional economies.* New York: Free Press.

Painter, D. (1999) Public–private partnerships for environmental improvement in Asia: The role of the US Agency for International Development. Paper presented at the Mayor's Asia Pacific Environment Summit, Hawaii, USA, 31 January–3 February.

Parivartan (2005) *Delhi Water Supply and Sewerage Project: An Analysis.* Available online at http://planningcommission.nic.in/data/ngo/csw/csw_5.pdf (accessed 11 July 2005).

Parivartan (2004) Public eye on public services: Delhi water reforms. Press Release. Available online at http://www.delhiwater.org/whoisrunningdelhijalboard.htm (accessed 22 August 2006).

Pasteur, K. (2001) Policy processes: What are they and how can they be influenced in support of sustainable livelihoods? Available online at: http://www.livelihoods.org/ (accessed 9 December 2007).

Patnaik, P. (2000) Economic policy and its political management in the current conjuncture. In F. Frankel, Z. Hasan, R. Bhargava and B. Arora (eds.) *Transforming India: Social and Political Dynamics of Democracy*. New Delhi: Oxford University Press, pp. 231–53.

Pedersen, D. J. (2000) Explaining economic liberalization in India: State and society perspectives. *World Development* 28 (2): 265–82.

Pitman, K. (2002) India: World Bank assistance for water resources management – A country assistance evaluation. Working Paper. Washington, DC: World Bank Operation Evaluation Department.

Planning Commission (2002a) *India: Assessment 2002 – Water and Sanitation: A WHO-UNICEF Sponsored Study.* New Delhi: Planning Commission, Government of India. Available online at http://planningcommission.gov.in/reports/genrep/wtrsani.pdf (accessed 21 October 2006).

Planning Commission (2002b) *Report of the PPP Sub-Group on Social Sector: Public Private Partnership.* New Delhi: Planning Commission, Government of India, November. Available online at: http://planningcommission.gov.in/reports/genrep/rep_ppp.pdf> (accessed 4 December 2005).

Planning Department (2008) *Economic Survey of Delhi 2007–2008.* New Delhi: Government of Delhi. Available online at http:delhi/planning.nic.in/Economic%20Survey/ES2007–08/C1.PDF (accessed 20 September 2008).

Planning Department (2004) *Economic Survey of Delhi 2003–2004.* New Delhi: Government of Delhi.

Planning Department (2002) *Economic Survey of Delhi 2001–2002.* New Delhi: Government of Delhi.

PPIAF – Public–Private Infrastructure Advisory Facility (2003) *Report on the Water Sector Policy Reform Initiative, India.* Washington, DC: PPIAF. Available online at http://wbln0018.worldbank.org/ppiaf/activity.nsf/files/Water+Sector+Policy+Reform+Initiative+India.pdf/$FILE/Water+Sector+Policy+Reform+Initiative+India.pdf (accessed 4 May 2006).

PwC – PricewaterhouseCoopers (2004) *Delhi Water Supply and Sewerage Project Preparation Study.* London: PwC.

Raghu (2005) Privatization of Delhi water supply. *People's Democracy* 29, 41, 9 October.

Rajamani, M. (2004) Financing municipalities and sub-national governments in India. Regional Round Table on Asia, 2nd Conference on Financing Municipalities and Subnational Governments. Washington, DC, 30 September.

Rashtriya Jal Biradari (2002) Open letter written to announce the launch of the campaign against the proposed water policy of the Government, 25 March.

Ratnagar, S. (2001) *Understanding Harappa: Civilization in the Greater Indus Valley.* New Delhi: Tulika.

Rao, V. M. (2001) The making of agricultural price policy: A review of CACP reports. *Indian School of Political Economy* 13 (1): 1–28.

Rao, V. M. (1999) Policymaking for agricultural development: Why it misses the poor. *The Indian Journal of Labour Economics* 42 (1): 49–57.

Ravindran, P. (2005) India's water economy – World Bank prescription does not hold water. *The Hindu-Business Line*, 11 October.

Rein, M. and Schon, D. (1993) Reframing policy discourse. In F. Fischer and J. Forestor (eds.) *The Argumentative Turn in Policy Analysis and Planning*. London: University College London Press, pp. 145–66.

RFSTE – Research Foundation for Science, Technology and Ecology (2006) *The Story of Delhi: Water Privatization vs. Water Democracy*. New Delhi: RFSTE/Water Workers Alliance.

RFSTE (2005) *Financing Water Crises*. Research Report. Water Sovereignty Series. New Delhi: RFSTE.

Roe, E. (1991) Development narratives, or making the best of blueprint development. *World Development* 29: 1677–94.

Rosenau, J. N. (2002) Governance in a new global order. In A. McGrew (ed.) *Governing Globalization: Power, Authority and Global Governance*. Cambridge: Polity Press, pp. 70–86.

Rosenau, J. N. (1997) *Along the Domestic Frontier: Exploring Governance in a Turbulent World*. Cambridge: Cambridge University Press.

Rudolph, L. and Rudolph, S. (2001) Redoing constitutional design: From an interventionist to a regulatory state. In A. Kohli (ed.) *The Success of India's Democracy*. Cambridge: Cambridge University Press, pp. 127–62.

Rudolph, L. and Rudolph, S. (1987) *In Pursuit of Lakshmi: The Political Economy of the Indian State*. Chicago: University of Chicago.

Ruet, J. (2002) Water supply and sanitation as "urban commons" in Indian metropolises: How redefining the state/municipalities relationships should combine global and local de facto "commoners." Paper presented at The Commons in an Age of Globalisation, the Ninth Conference of the International Association for the Study of Common Property, Victoria Falls, Zimbabwe, 17–21 June.

Ruet J. and Zérah, M. H. (2001) *Water Supply and Sanitation in Indian Metros:Bombay, Calcutta, Madras*. Research Report. New Delhi: Centre de Sciences Humaines.

Sabatier, P. (1987) Knowledge, policy-oriented learning and policy change. *Knowledge: Creation, Diffusion, Innovation* 8, June: 649–92.

Sabatier, P. (1986) Top-down and bottom-up approaches to implementation research: A critical analysis and suggested synthesis. *Journal of Public Policy* 6: 21–48.

Sabatier, P. and Jenkins-Smith H. (1993) (eds.) *Policy Change and Learning: An Advocacy Coalition Approach*. Boulder: Westview.

Sabatier, P. and Jenkins-Smith, H. (1988) (eds.) Symposium issue on policy change and policy-oriented learning: Exploring an advocacy coalition framework. *Policy Sciences* 21: 123–278.

Sabatier, P., Vedlitz, A., Focht, W., Lubell, M., Matlock, M. (2005) (eds.) *Swimming Upstream: Collaborative Approaches to Watershed Management*. Cambridge, MA: MIT Press.

SANDRP – South Asia Network on Dams, Rivers and People (2002) *Update on Dams Options and Related Issues*. New Delhi: SANDRP.

Sassen, S. (1996) *Losing Control? Sovereignty in an Age of Globalization: The 1995 Columbia University Leonard Hastings Schoff Memorial Lectures*. New York: Columbia University Press.

Satya, L. (2001) Water control and its consequences in colonial and post colonial India. Paper presented at the European Conference on Modern South Asian Studies, 20–2 September.

Schön, D. (1983) *The Reflective Practitioner: How Professionals Think in Action*. New York: Basic Books.

Schonwalder, G. (1997) New democratic spaces at the grassroots? Popular participation and Latin American local governments. *Development and Change* 28 (4): 753–0.

Schram, S. (1993) Postmodern policy analysis: Discourse and identity in welfare policy. *Policy Sciences* 26: 249–70.

Scoones, I. (2003a) Making policy in the "new economy": The case of biotechnology in Karnataka, India. IDS Working Paper No. 196. Sussex, UK: IDS.

Scoones, I. (2003b) Regulatory manoeuvres: Bt cotton in Karnataka, India. IDS Working Paper 197. Biotechnology Policy Series 14. Sussex, UK: IDS.

Seetaprabhu, K. S. (2001) *Economic Reform and Social Sector Development: A Study of Two Indian States*. New Delhi: Sage.

Sehgal, R. (2007) Reclaiming public water: The experience of Delhi. Available online at http://www.tni.org/books/waterdelhisehgal.pdf (accessed 23 May 2006).

Sethi, A. (2005) Delhi's pipedream. *Frontline* 22 (17), 13–26 August. Available online at http://www.hindu.com/fline/fl2217/stories/20050826003903100.htm (accessed 27 February 2009).

Shackley, S. and Wynne. B. (1995) Global climate change: The mutual construction of an emergent science–policy domain. *Science and Public Policy* 22: 218–30.

Sharma, P. N. (ed.) (1998) *Ripples of Society: People's Movements in Watershed Development*. New Delhi: India Gandhi Peace Foundation.

Sharma, S. (2005) Privatization of water supply: There is a hole in the bucket. Available online at: http://www.infochangeindia.org/analysis81.jsp (accessed 24 May 2006).

Sharma, S., Naqvi, A. and Zafri, A. (2004) *Delhi Jal Board Financial Sustainability is Possible Through Public–Public Partnership*. Research Report. New Delhi: RFSTE/ Water Workers Alliance, 8 September.

Shastri, V. (2001) The politics of economic liberalization in India. In J. L. Bajaj (ed.) *The Indian State in Transition*. New Delhi: National Council of Applied Economic Research, pp. 1–33.

Shiva, V. (2005) Is privatisation and water pricing a sensible policy? Community rights are vital for democracy and ecology. *Financial Express*, 22 March.

Shiva, V. (2005) Response of Dr V. Shiva to the CEO of the DJB, Mr Rakesh Mohan on the dialogue on Delhi's water privatization. In RFSTE, *Privatization of Water*. Research Report. New Delhi: RFSTE.

Shiva, V. (2005) Water privatisation and water wars. 14 July. Available online at: http://www.zmag.org/Sustainers/Content/2005-07/12shiva.cfm (accessed 12 December 2005).

Shiva, V. (2003a) Killing the Ganga. *The Hindu*, 31 August.

Shiva, V. (2003b) Captive waters. Available online at: http://www.resurgence.org/resurgence/issues/shiva219.htm (accessed 11 October 2005).

Shiva, V. (2002) *Water Wars: Privatization, Pollution and Profit*. Cambridge, MA: South End Press.

Shiva, V., Bhar, R. H., Jafri, A. H. and Jalees, K. (2002) *Corporate Hijack of Water: How World Bank, IMF and GATS-WTO Rules are Forcing Water Privatization*. New Delhi: RFSTE.

Shiva, V. and Jalees, K. (2003) *Ganga: Common Heritage or Corporate Commodity*. New Delhi: RFSTE.

Shore, C. and Wright, S. (1997) Policy: A new field of anthropology. In C. Shore and S. Wright (eds.) *Anthropology of Policy: Critical Perspectives on Governance and Power*. London: Routledge, pp. 3–37.

Simon, H. A. (1957) *Models of Man*. New York: Wiley.

Singh A. K. (2005) Thirty liters for some, 1,600 for others: Inequalities in Delhi's water supply. Available online at http://www.infochangeindia.org/agenda3_04.sp (accessed 30 August 2007).

Singh, A. K. (2004) Global freshwater crisis: Privatization of rivers in India. *Vikas Adhyayana Kendra*, June: 16.

Singh. M. (1997) Interviews: Liberalization and globalization: Where is India heading? *World Affairs* 1 (1): 16–42.

Smith, G. and May, D. 1980. The artificial debate between rationalist and incrementalist models of decision making. *Policy and Politics*. 8: 147–61.

Smith, M. (1993a) *Pressure, Power and Policy: State Autonomy and Policy Networks in Britain and the United States*. London: Harvester Wheatsheaf.

Smith, M. (1993b) *Pressure, Power and Policy*. Hemel Hemstead, UK: Harvester Wheatsheaf.

Sreedhar, G. and Ravindra B. N. (2006) Economics of micro irrigation systems. In K. N. Rao (ed.) *Water Resources Management: Realities and Challenges*. New Delhi: New Century, Ch. 13.

Standing Committee on Urban and Rural Development (2002) *Twenty-Eighth Report*. New Delhi: Ministry of Rural Development.

Strange, S. (1996) *The Retreat of the State: The Diffusion of Power in the World Economy*. Cambridge, NY: Cambridge University Press.

Sudan, R. (2000) Towards SMART government: The Andhra Pradesh experience. *Indian Journal of Public Administration* 46 (3): 401–11.

Thakkar, H. (2004) A national policy. *Water Politics: The People's Movement* 1 (5), September–October.

Thapar R. (1966) *A History of India*. Vol. 1. London: Penguin.

The Economic Times (2003) Water Man speaks, 12 June.

The Economist (2005) Private worries: The water industry in India, 13 August.

The Hindu (2005) Delhites gearing up to protest against water privatization, 18 October.

The Hindu (2005) Khurana warns of public movement over water, 21 April.

The Hindu (2005) NGOs willing to run Sonia Vihar Plant, 6 April. Available online at http://www.hindu.com/2005/04/06/stories/2005040616700300.htm (accessed 22 November 2005).

The Hindu (2005) Privatization a bonanza for water companies, 13 July.

The Hindu (2005) Protest against water privatization, 12 July.

The Hindu (2005) Sheila accused of misleading assembly, 10 April.

The Hindu (2004) A front against water privatization, 23 March.

The Hindu (2004) No privatization of DJB: Sheila, 2 October.

The Hindu (2004) Sonia dedicates Dhaula Kuan flyover to the people of Delhi 20 November. Available online at http://www.hindu.com/2004/11/20/stories/2004112005780400.htm (accessed 6 November 2006).

The Hindu (2003) Privatization will precipitate water crisis, 19 December.

The Times of India (2005) Sonia Vihar a tough jinx to crack, 17 June.

The Times of India (1998) Swadeshi liberalism, 15 April.

The Times of India (1998) 21 March.

The Times of India (1998) 26 March.

The Tribune (2008) Congress Starts rallies; Sheila lists achievements, 14 November.

The Tribune (2007) Conserve water to battle scarcity, stresses Sheila, 14 April.

The Tribune (2005) Hundreds of objections to Master Plan for Delhi 2021, 15 July.

The Tribune (2002) Delhi to get 20 percent more water, 22 June.

Thomson, A. (2000) Sustainable livelihood approaches at the policy level. Paper presented at the FAO E–conference and Forum on Operational Ways of Applying a Sustainable Livelihoods Approach.

Thompson, W. (1961) *Modern Organization*. New York: Knopf.

Upreti, B. C. (1993) *Politics of Himalayan Rivers: An Analysis of the River Water Issues of Nepal, India and Bangladesh*. Jaipur: Nirala.

Vaidyanathan, A. (1999) *Irrigation Management in India: Role of Institutions.* Oxford University Press, New Delhi.

Varshney, A. 1989. Ideas, interest and institutions in policy change: Transformation of India's agricultural strategy in the mid-1960s. *Policy Sciences* 22 (3–4): 289–323.

Viswanathan, S. (2003) Soft drinks and hard choices. *Frontline,* 20 June.

Vombatkere, S. (2005) National water policy 2002, A critique. In G. Budhya (ed.) *Implementation of ADB Water Policy in Karnataka Urban Infrastructure Development and Coastal Environment Management Project (KUDCEMP), South India.* Bangalore: Urban Research Centre. Available online at http://www.adb.org/Water/Policy/consultations/NGO-FORUM-IND (accessed 20 December 2005).

Waterbury, J. (1990) The heart of the matter? Public enterprise and the adjustment process. In S. Haggard and R. Kaufmann (eds.) *The Politics of Economic Adjustment: International Constraints, Distributive Politics and the State.* Princeton: Princeton University Press, pp. 182–217.

Watkins, T. (Year not known) The economic history of the city of Delhi, old and new. Available online at: http://www.sjsu.edu/faculty/watkins/delhi2.htm (accessed 27 January 2006).

Weber, M. (1991) Bureaucracy. In H. Gerth and C. W. Mills (trans. and eds.) *From Max Weber: Essays in Sociology.* London: Routledge, pp. 159–264.

WEF – World Economic Forum (2005) Business alliance on water launched to create public–private water projects. Press Release. New Delhi: World Economic Forum, 29 November. Available online at: http://www2.weforum.org/site/homepublic.nsf/Content/Business+Alliance+on+Water+Launched+to+Create+Public-Private+Water+Projects.html (accessed 27 February 2009).

Weiner, M. (1999) The regionalization of Indian politics and its implications for economic reform. In A. H. Jeffrey, D. Sachs, A. Varshney and N. Bajpai (eds.) *India in the Era of Economic Reforms.* New Delhi: Oxford University Press, pp. 261–95.

Weiss, C. (1986) Research and policymaking: a limited partnership. In F. Heller (ed.) *The Use and Abuse of Social Science.* London: Sage, pp. 214–35.

Weiss, L. (1998) *The Myth of a Powerless State.* New York: Cornell University Press.

Wilks, S. and Wright M. (1987) *Comparative Government-Industry Relations.* Clarendon Press: Oxford.

Wilson, D. (1993) Tourism, public policy and the image of Northern Ireland since the troubles in Ireland. In B. O. Connor and M. Cronon (eds.) *Tourism in Ireland: A Critical Analysis.* Cork: Cork University Press.

Wolf, A. T. (2003) Conflict and cooperation: Survey of the past and reflection for the future. Paper prepared for the UNESCO/Green Cross International Program: "From Potential Conflict to Cooperation Potential: Water for Peace" in collaboration with the Organization for Security and Cooperation in Europe. Available online at: http://webworld.unesco.org/water/wwap/pccp/cd/pdf/history_future_shared_water_resources/survey_water_conflicts_cooperation.pdf (accessed 23 November 2005).

World Bank (1994) *World Development Report 1994: Investing in Infrastructure.* New York: Oxford University Press.

World Bank (1993) Water resources management. Policy Paper. International Bank for Reconstruction and Development. Washington, DC: World Bank.

Zald, M. (1996) Culture, ideology, and strategic framing. In D. McAdam, D. J. McCarthy and M. Zald (eds.) *Comparative Perspectives on Social Movements: Political Opportunities, Mobilizing Structures and Cultural Framings.* Cambridge, UK: Cambridge University Press, pp. 261–74.

Zérah, M. H. (2001) *The Cancellation of the Pune Water Supply and Sewerage Project.* New Delhi: Water and Sanitation Program, South Asia.

Zérah, M. H. (2000a) Household strategies for coping with unreliable water supplies: The case of Delhi. *Habitat International* 24 (3): 295–307.

Zérah, M.H. (2000b) Water: Unreliable Supply in Delhi. New Delhi: Manohar.

Zérah, M. H. and Llorente, M. (2003) Urban water sector: Formal versus informal suppliers in India. *Urban India* 22 (1): 1–15.

Index

ACF (Advocacy Coalition Framework), 15–16, 19, 148
actors, 2–3, 6–9, 12–18, 20, 23–4, 26–7, 45, 96, 117–18, 137, 139–41, 143, 145, 147–8, 150–1, 153; complex configuration of, 26, 117; constellation of, 3, 13, 150; external, 136; grassroots, 25; key, 146, 149; local, 3; mainstream, 99; multiple, 8, 13; networks of, 6, 12, 14, 144, 150; non-state, 7; political, 19, 35; position, 18; positioned, 152; social, 2
ADB (Asian Development Bank), 24, 63–4, 68–9, 127, 129, 154
ADB's Water Policy, 64
administration, 33, 41, 90
administrative set-up, 49
administrators, 114, 140; subordinate, 11
advisors, 35–7; economic, 22, 31, 33, 37
advocacy coalitions, 15, 19, 26; competing, 15
advocacy efforts, 57
Agarwal, 24, 38, 154
agencies, 2, 5, 9, 12–14, 26, 64, 67–9, 117, 137, 147; benevolent, 2; bilateral development, 104; development aid, 21; donor, 132, 141; external, 23, 28, 66; external support, 65; independent, 42; international, 5, 67–9; international financial, 38, 55; multidonor, 55; nongovernmental, 63; political theory, 20

agenda setting, 7, 9–12, 147, 149
agendas, 1, 10, 12, 20, 33, 58, 72, 104, 115, 118; framing set policy, 17, 150; mainstream, 117; national, 52, 58, 66; political, 27, 53; setting policy, 13; visionary, 104
agents, 12; lateral, 33; private, 43, 51
agitation, 6, 30, 95; year-long, 90
agreement, 61, 67–8, 101, 115, 124; fair, 125; interstate, 76
agricultural policy, 23
agricultural productivity, 46
agriculture, 25, 28, 47
Ahluwalia, 32, 36–8, 154
Alexander, 31
algae, 88
Ali-Bogaert, 25, 154
Allchin, 45, 154
alliances, 14–15, 22, 29, 42, 47, 63, 93, 98, 113, 118, 120, 132, 137, 150; building, 89; single national, 95
allocation, 48, 50; equal, 131; risk, 129
Allouche, 2, 4, 65, 97
amendment, 49
analysis: article, 22; centered discursive, 150; class, 150; empirical, 16; final, 146; interface, 150; narrative, 17, 150; structural, 12
analysts, 3; global process, 3
ancient, 45
Anderson, 10, 154
Andhra Pradesh, 23–4, 68–9
anti-democratic, 137
anti-globalization debate, 141

approaches, 2, 4, 9, 11, 13–15, 17–18, 26, 32, 41, 50, 54, 66, 69, 103, 107, 149–52; centered, 141, 150–1; incremental, 41; multidisciplinary, 106; orthodox, 142; participatory, 59, 106, 116, 142; standard, 37; supply-side, 61; wait-and-see, 38
Apthorpe, 16–17, 150
Aquazur, 88
areas: district-metering, 125; environmental, 111; marshy, 85; planned, 79; rural, 47, 49, 81, 130–1, 138; total, 47; urban, 47, 49, 69
Arjun Sengupta, 31
Arora, Gopi, 1, 21, 32, 154
Asian Development Bank *see* ADB
Assam, 30
assessment, 28, 41; environmental, 90; titled India, 61
assets, 100, 113; commercial, 43; natural, 46
assistance: excessive loan, 109; external, 49; project-by-project, 54; technological, 68
Atkinson, M.M., 12, 154
augmentation, 47, 82, 8
authorities, 3, 10, 29, 45, 51, 66, 82, 118; cede, 3; legitimate, 10; local, 51
autonomy, 31; commercial, 73; fiscal, 70; local, 46; political, 45; temporary, 33

Babbie, E., 7, 154
back door privatization, 95, 113, 133
Barker, C., 11, 154
Barve, V., 69
Basham, A.L., 70
beliefs, 15, 35, 120, 134
beneficiaries, 111, 131; intended, 102, 110, 124
Benford, R.D., 14
best practices, 57, 63, 104, 108; corporate, 63
Bhaduri, A., 22, 42
Bhagirathi, 77, 83
Bhakra Nangal Dam, 48
Bhartiya Jagriti Mission, 95, 97
Bhartiya Janata Party: *see* BJP; nationalist, 31

Bhartiya Kisan Union, 95
Bhattacharya, S., 46
bills, 42–3, 67, 93, 128; country's import, 30; growing import, 30; monthly water, 130
BJP (Bhartiya Janata Party), 31, 38, 42–3, 144
BMZ, 68
boards, 49, 52; state electricity, 43
bonus, 101, 105
BOO (Build-Own-Operate), 62
BOT (Build-Operate-Transfer), 62, 94, 97, 105, 112, 119, 124
bottom up, 142
Boulding, 25
Brady, 154
Braybrooke, D., 13
Briscoe, John 54, 64
budget, 15, 40–1, 66, 141; limited, 32; total O&M, 12
Build-Neglect-Rebuild, 64
Build-Operate-Transfer *see* BOT
Build-Own-Operate *see* BOO
bureaucracy, 5, 10, 21, 33, 35, 58, 66, 98, 104, 112–14, 140, 143, 148–9; efficient state, 22; sector's, 55; unreconstructed colonial, 5
bureaucratic elite, 34, 40, 60, 62; lateral, 34; like-minded, 29
bureaucrats, 31, 34–5, 40, 66, 108–9, 112, 144, 146
business, 20, 24, 31, 39–40, 63, 123, 148, 154
business model, 62

Callon, M., 14
Camdessus, Michael, 38
campaigns, 42, 95, 113, 120–1
canals, 46–8, 125
Canepa, C., 79
capacity, 3, 36, 55, 64, 67, 77, 82–3, 87, 125; advisory, 35; half, 131; in-depth clogging, 88; limited, 109; outgrowing, 75; state's, 151; technical, 74; transformative, 3
capacity building, 74
capital, 2, 30, 39, 41, 46–7, 65, 71, 92, 112, 119, 130–1, 134, 138; global, 143; local, 122; national, 90; nation's, 100

capital costs, 60
capital expenditures, 40, 96
capital investments, 46, 91, 97, 103,
    142; involved Ondeo Degremont's,
    126
capital markets, 55; global, 36
Carter, Michael, 100–1, 103, 140
causal model, 15; empirical, 16
CEO (chief executive officer), 107, 109,
    124
CFWD (Citizens Front for Water
    Democracy), 93–5, 97, 113, 118,
    120–2, 124, 128–30, 132–6, 139,
    143–5, 147, 153
CGWB, 77
Chakrabarty, 31
Chandra, Bipin 119
change agents, 143
Change Team, 29, 33–4; pivotal, 34
Chidambaram, 40
chief executive officer *see* CEO
Chief Minister, 93, 113; Chief Minister
    of Delhi, 72, 94, 97, 105, 109, 112
CII (Confederation of Indian Industry),
    39, 63
citizens, 3, 27, 72, 87, 89, 93, 100,
    102, 113, 119, 121, 132, 135–7, 141,
    147–8
city, 25, 62, 69, 71–2, 74–5, 77–8, 81,
    90–1, 95–6, 100, 102, 105, 109, 112,
    124, 142; important, 71; largest,
    75; livable, 72; post-colonial, 71;
    selected, 57
civil disobedience, 93
civil society, 14, 27, 57, 60, 66, 93, 115,
    118, 137, 141, 146–8, 151–2
claims, 14–15, 26, 63, 96, 100, 109, 111,
    115, 118–20, 122, 126–8, 134, 137,
    141, 143, 145; raised, 122; religious,
    133; reported, 128; staked, 141
Clarke, N., 11
Clay, E., 12, 14
coalition formation, 16
coalition parties, 43
coalitions, 15–16, 18, 20, 42, 44, 60,
    118, 150; discursive, 150; unstable,
    52
Cobb, R., 13, 34
coercion, 27, 40

Colebatch, H.K., 10, 14
Coleman, 12, 154
Collins, A., 118
colonial India, 47
colonial rule, 30, 46, 70
colonies, 81, 130; congested, 81;
    posh, 131; posh South Delhi, 131;
    regularized, 81; resettlement, 110,
    131; residential, 95; settlement, 110;
    unauthorized, 80, 131, 138
command and control, 5
commercialization, 24, 37, 41, 46,
    51–3, 62, 68; Commercialization of
    infrastructure projects, 52
commission, regulatory, 43, 89
committees, 24, 30, 36–7, 41, 51–3,
    89; appointed, 21; expert, 51; high
    powered, 41
commodification, 134
communities, 2, 14, 45, 59, 63, 111, 121,
    136, 139, 154; donor, 115; epistemic,
    12–13, 149; farmer, 145; local, 95,
    132; low income, 92; lucrative, 131;
    rural, 88; spiritual, 141
community rights, 59, 120
companies, 7, 46–7, 89, 91, 93, 97,
    101–3, 108, 116, 122–3, 125–6,
    134; based, 69; international, 122;
    managing, 91; multinational, 128
conditionalities, 41, 66; economic, 115;
    fiscal, 68; institutional, 24
conflicts, 48, 116, 128, 145; dam, 25
Congress Party, 30, 33, 36, 38, 42, 144;
    ruling, 38
connections, 22, 27; globalized, 144;
    group meter, 128; piped, 122, 129
consensus, 14, 19, 32, 35, 39, 41, 57, 62,
    66, 104–5, 115, 118; political, 41
Constitution of India, 48
construction, 17, 48, 84, 87–8, 97,
    103, 124, 143, 150; changing, 143;
    discursive, 99; horizontal, 151;
    mutual, 12; private, 46–7; railway, 46;
    social, 28; vertical, 151
consultants, 68–9, 90, 92, 96–7, 104,
    114–15, 122, 128, 130, 132, 139–40,
    144; expert, 126
consultants PricewaterhouseCoopers ,
    PwC, 114

consultation, 60, 115, 118, 141–2; nationwide, 60; public, 103
consumers, 52, 79, 81, 91, 102, 105, 109–10, 125–6, 130
consumption: domestic, 131; energy, 92; rationalize, 52; wasteful, 123
contamination, 62, 106, 111
contract, 7, 85, 87–91, 94, 97, 101, 105, 108, 119, 123, 125, 133, 135–7, 147, 151; consultancy, 97; million, 122
Corbridge, 22
corporations, 4, 49, 88, 120, 123, 126, 134, 145; corrupt transnational, 134; domestic, 68; global, 125; global water, 127; municipal, 49, 68–9; private, 90, 121, 133; thirsty, 133; transnational, 60
corporatization, 62, 127
cost effectiveness, 68
cost efficiency, 74
cost recovery, 51–2, 54, 69, 79, 92, 103, 109, 112, 122, 130; full, 127; low, 51, 109
costs, 51, 54, 60, 65, 85, 87–8, 96, 101–2, 105, 109–11, 113, 123–4, 126, 129–30, 135–7; energy, 92; environmental, 62, 88; estimated, 82; financial, 88; fixed, 50; high, 51, 109, 111, 123; lower project, 52; operating, 79, 102; total operational, 110
Cotton, 46–7
CPCB, 88
CPI, 38
crisis, 22, 39, 58, 65, 67, 69, 154; acute financial, 46; atmosphere of, 114, 140; fiscal, 29
critique, 9, 11, 22; contemporary, 3; environmental, 25; gender, 25
Crow, 24
CSE, 76–7, 79, 131
culture, 4, 55, 65, 96, 134–5, 137–8, 141; intellectual, 148; living, 89, 119
Currie, 23

Daga Shivani, 81
Dandavate, Madhu, 37
Dash, K., 30–2, 38–40
Dearing, J.W., 14

debt, 70, 97, 138; external, 38; high public, 4
debt relief, 53
debt-service ratio, 38
Deccan Plateau and Thar Desert, 75
decentralization, 29, 49, 52, 62, 65, 151
decisions, 2–4, 10–13, 30, 39, 48, 59–60, 89, 96, 105–6, 118, 133, 149–50, 152; authorized, 10; bureaucratic, 10; conscious, 108; democratic, 132; determinate, 11; influence project, 146; investment, 100; national, 53; real economic, 31; tariff-setting, 109; transferring, 51
Dehat Morcha, 85, 88, 95
Deleon, P., 149
Delhi, 7–8, 26–7, 63, 69, 71–83, 85–90, 93–7, 99–100, 104–6, 109–14, 116, 118–21, 123–4, 126–34, 139–49, 151; citizens of, 100, 102, 136; city of, 71; entered, 93; situates, 69; state of, 6, 115, 132, 152; transform, 71; urban water reform policy uses, 9; vision of, 75, 107; water problems of, 96, 103
Delhi Urban Environment and Infrastructure Improvement Project, 72
Delhi Water Sewer and Sewage Disposal Employees Union, 95
Delhi Water Supply and Sewage Disposal Undertaking, 77
Delhi Water Supply and Sewage Sector Reform Project, 94
Delhi Water Supply and Sewerage Policy, 119, 122
Delhi Water Supply and Sewerage Project, 97
Delhi Water Supply and Sewerage Project Preparation Study, 90
Delhi Water Supply and Sewerage Reform Project, 116
Delhi Water Supply and Sewerage Sector Project, 90
Delhi Water Supply and Sewerage Sector Reform Project *see* DWSSSRP
delivery projections, 78
delivery services, 100; efficient, 78
delivery system, 108
demand management, 51, 110–11

demand supply gap, 78
democracy, 22, 37, 118, 132–3;
    functional, 148; participative, 21;
    political, 41; subversion of, 133
depoliticization, 27
Desarda, 138
DEUIIP, 90
devolution, 65, 107
Devraj, R., 59
DFID (Department for International
    Development), 21, 68
Dharmadhikary, 129
dimensions, 100; cultural, 58; empirical
    cognitive, 19; horizontal, 151;
    relational, 150
discourse analysis, 17, 150
discourse coalitions, 12, 18–19, 104,
    153
discourse literature, 19
discourses, 7–8, 10, 16–20, 26, 33, 35,
    53, 58, 65, 99, 102–3, 114, 118, 137,
    141–3, 150–1; alternative, 103, 115,
    142, 145; competing, 26, 139, 141;
    expert, 16; global, 1; hegemonic,
    115, 118; multiple, 20; nationalist–
    protectionist, 112; pervasive, 8;
    political, 18, 27; technological, 143;
    variant, 27
discourses of power, 99, 115, 140, 142
discursive approaches, 12
discursive package, 103
distribution, 47–50, 52, 71, 77, 82, 87,
    89–90, 106, 112, 136, 146; bulk, 82;
    equitable, 132; inequitable, 79, 129,
    131; retail, 82; unequal, 131
distribution zones, 90
district-metering areas (DMAs), 125
Dixit, 24; Dixit, Sheila, 72, 96–7, 105,
    112, 146
DJB (Delhi Jal Board), 74–5, 77–9,
    81–2, 87–102, 104–14, 116–17,
    123–6, 129–30, 136–8, 142, 148
DMAs (district-metering areas), 125
Dobuzinskis, L., 13
donor–recipient relationship, 114, 140
donors, 5, 14, 21, 25, 27, 69, 115, 117,
    146–7, 151; international, 6, 20;
    largest, 54
Dreyfus, H.L., 18

drinking water, 68, 102, 111, 121, 131;
    safe, 107
Dror, Y., 13
Dryzek, J., 17
Dubey, Muchkund 119
DUEIIP, 72–80, 87
DWSSSRP (Delhi Water Supply and
    Sewerage Sector Reform Project), 91,
    94, 104, 110–11, 115, 128, 142

Echeverri-Gent, J., 23
Ecology, 94, 130
economic crisis, 31, 39–40; developing,
    37
economic liberalization, 5, 22, 27,
    29–44, 50
economic nationalism, 36, 42
economic policy, 29, 32–3, 39; friendly,
    31
economic policy initiatives, 30
economics, 24, 37, 154; urban, 37
economies, 33, 39–40, 43, 50, 52–3,
    67, 100; agricultural, 43; command,
    43; competitive, 39; growing, 64;
    international, 2; national, 42; new,
    142
efficiency, 26–7, 32, 36, 51–2, 55,
    63, 65–6, 68, 74, 95, 100, 107–8,
    112, 115, 121, 127; economic,
    65–6; improving water management,
    63; increased, 36; increasing, 73;
    operational, 107
elections, 42, 44
electoral politics, volatile, 146
electoral victories, 31, 38
Elhance, A.L., 24
elite, 33, 35, 40, 42, 65, 138; business–
    science, 143; civil service, 21
embankments, 48
Emerson, R., 7
entrants, 75; lateral, 35
entrepreneur, 39
era, 30, 45, 49; economic, 23; neoliberal,
    4, 8, 141; post-Westphalian, 3;
    pre-reform, 142
Etzioni, A., 13
exit policy, 41
expertise, 6, 19, 101, 103, 108–9, 112,
    122, 129, 142, 144; global, 114, 140;

managerial, 108; operating, 101; private, 107, 112; technical, 102; technological, 109, 140, 142

experts, 60, 88, 97, 103–4, 109, 126, 142, 146, 153; global, 140; network of, 114, 140; technical, 114, 140

feasibility, 122; technical, 111
Finance Minister, 36–8
Finance Minister Man Mohan Singh, 40
Finance Minister Yashwant Sinha, 42
Finger, 2, 4, 65, 97
Fischer, 15–17, 19–20
Forster, 17
Foucauldian, 18, 109
Foucault, M., 17–18, 27
framework: advocacy coalition, 15, 148; conceptual, 9; neoliberal, 58; regulatory, 51; strategic, 72; well-enforced legal, 12
framing, 6, 9, 17, 19, 35, 118, 142, 150
Frankel, F., 23
free market doctrine, 146
free market economics, 121
French MNC Ondeo Degremont, 7
French MNC Suez in India, 122
French multinational Suez, 85

Gandhi, Indira, 30–1, 40, 67, 69, 79, 96, 103
Gandhi, Rajiv, 31–3, 35, 40, 49, 67, 146
Gandhi, Sonia, 72, 94, 119, 123, 132
Ganeshan, 35
Ganga, 87, 89, 94, 97, 119, 124, 133–4, 145; mother, 134; sacred, 119, 124, 145
Ganga Canal, 77, 83, 85
Ganga River, 83–4, 88–9, 133
Ganga Yatras, 89, 94, 119
Ganges River Valley, 75
Gangotri Glacier, 83
Gasper, D., 16–17
GATS (General Agreement on Trade in Services), 53
GATT (General Agreement on Tariffs and Trade), 127
GDP (gross domestic product), 38

Gerth, H., 11
Giddens, A., 2
Gilmartin, D., 47
GkW, 128
globalization, 1–4, 7, 29, 42, 52–3, 65–6; age of, 2, 4; defined, 2; economic, 4; term, 2
Goldman, M., 58
Gordon, I., 12
governance, 1, 3–4, 18, 23, 27, 29, 52, 61, 65, 71, 103, 111–12, 114, 140, 154; corporate, 90; financial, 3; transparent, 132
Government, 11, 32–3, 35–9, 42, 44, 46–7, 49, 54, 61–3, 66, 71–2, 82, 104–5, 109–14, 119–20, 135–6; central, 5, 31, 37, 60–1, 67, 115; coalition, 42; local, 148; secular, 42
Grillo, R., 17
Grindle, M., 12, 20
gross domestic product *see* GDP
Groundwater Regulation, 56
Grove, R., 154
growth, 4, 20, 30, 37, 51, 103, 107, 114, 121, 140, 149; decadal, 75; potential, 72; rapid, 36; sustainable, 54; uncontrolled, 77
Gulati, A., 25
Gunn, 11
Gupta, 77
Gyawali, D., 24

Haas, P., 12–13, 150
Habermas, J., 4
Haggard, S., 35, 40
Hajer, Martin, 12, 17–20, 104
Hardwar Declaration, 134
Harriss, B., 21–3
Hegde, R., 42
Hjern, B., 13
Hogwood, B., 11
Horowitz, D., 20
households, 25, 79, 101, 111, 116, 123, 125
Houtzanger, 11, 118
Howlett, M., 10, 13
hydro-apartheid, 120
hydrocracies, 6
hydroelectric projects, 48

hydropolitics, 24; domestic, 6
hydropower, 49–50

IBAW (Indian Business Alliance on
  Water), 63
identities, 18
IFIs, 40–1
IIM (Indian Institute of Management),
  27, 138
IIT (Indian Institute of Technology), 138
ILFS (Infrastructure Leasing and
  Financial Services), 56, 68
IMF (International Monetary Fund), 4,
  6, 22, 30–3, 35–41, 53, 66, 127
Imperial Gazette of India, 46–7
incentives, 24, 55, 89, 125–6, 129, 131,
  146; based, 89, 125; claim, 125; fiscal,
  61; inadequate, 96, 103; productivity,
  88; special, 61
India, 1–2, 6, 8–9, 21–7, 29–32, 35–43,
  45–51, 53–8, 61–7, 70–2, 95–7,
  107–9, 121–4, 134, 144–8, 153–4;
  ancient, 45; central Government of,
  104; committed, 61; economic
  reform era of, 1, 21; independence,
  4; independent, 48; post-1985 period,
  32; post-colonial, 47; post-reform, 1;
  pre-British, 46; quit, 124; shining, 30;
  transforming, 43
indicators, 19, 43; positive, 114;
  quantifiable, 19; quantitative, 19
inequality in Delhi water supply, 80
inequity, 131, 133
inflation, 32
infrastructure, 4, 36, 47, 54–5, 57, 64,
  67–8, 75, 88, 90, 96–7, 114, 125–6,
  140, 142, 146; physical, 103; private,
  56, 154; trunk WSS, 92
infrastructure development, 36–7, 55,
  64, 104, 112
infrastructure projects, 52; urban, 61
infrastructure services, 52
institutions, 10, 15, 17, 24–5, 35, 40–1,
  45, 51, 60, 62, 79, 117, 141
interactions, 7, 12–13, 15, 17, 72,
  114, 147–8, 150–2; competing, 7;
  complex, 140; constant, 4; intense,
  139; micro level, 14; political, 20;
  repeated, 27

intervention, 42, 99, 110, 128–9, 143,
  146; critical, 90; international, 135;
  minimal, 48; moments of, 117, 120;
  potential areas of, 1; technical, 110
investments, 2, 52, 55, 61, 64, 91, 122,
  126, 128; donor, 147; estimated, 61;
  foreign, 39, 42, 44, 61, 96, 112, 146;
  private, 66; project-by-project, 54;
  public, 124
irrigation, private, 47
irrigation development, 25
irrigation management, 24–5
irrigation policy, 25
irrigation projects, 46
irrigation purposes, 47
irrigation works, 46; public, 46;
  sponsored, 46
Iyer, R., 24

Jafri, Afsar, 88
Jal Swaraj, 133
Jal Swaraj Yatra, 89, 94, 119
Janata Dal, 38
Jayal, Niraja, 23
Jenkins, R., 11, 22
Jenkins-Smith, 15, 19
Jessop, B., 3
Jha, 31, 70
JJ (Jhuggi Jhopri), 80, 100–1, 110, 116,
  128, 130
Joshi, Prabhat 119

Kahler, M., 40
Kanyinga, K., 118
Kaplan, 10
Karnataka, 68–9
Kaufman, 40
Kaur, N., 88
Kaviraj, S., 5
Keeley, J., 10, 12, 14, 20, 33
Kejriwal Arvind, 126
Keynesian, 64
Khagram, S., 25
Khurana, Madanlal, 24, 154
Kingdon, J., 13, 34
Kling, E.H., 15
knowledge, 9–10, 12, 17–18, 23, 26–7,
  29, 35, 57, 66, 99, 103–4, 115, 117,
  140, 143–4, 150–3; causal, 15;

cluster, 18; interplay of, 16, 26; local, 1, 141; particular political economy of, 8, 152; valid, 26
Kochanek, S., 22
Kohli, Atul, 22–3, 31–2, 35, 154
Kosambi, D., 70
kudimaramath, 45
Kumarangalam's, 42

Lal, Sumir, 102
Lasswell, Harold, 10
Latour, B., 14
Left Front, 42
Lele, Sharadchandra, 66
levels, 6, 11, 15, 61, 67, 101, 114, 130, 136, 148, 151; all-Delhi, 126; consumer, 24; differential, 52; global, 53; increasing ground water, 136; international, 53, 146; interstate, 24; local, 69, 145, 152; lower, 108, 112; managerial, 108; multiple, 7; national, 53, 66; optimum, 114; sub-state, 49
Levitt, Theodore, 2
Lewis, P., 2
Li, T., 65
liberalization, 5, 22, 30, 32–5, 37, 39–43, 52–3, 66, 70, 96, 152; external, 38; import, 39; progressive, 53
liberalization process, 22, 34–5, 37–9, 45–55, 57–70
Lindblom, C.E., 12–13
linear model, 2, 9, 11–12, 149
Lipsky, 11, 27
literature: end-of-the-state, 3; public administration, 2; social science, 2; varied, 12
livelihoods, 3
Llorente, M., 81
local bodies, 64; urban, 49, 107
Lok Sabha, 31, 44

management, 7, 11–12, 24–5, 32, 41, 47–9, 55–6, 58–9, 61, 64–5, 90–1, 97, 100, 105–8, 119, 136; bad, 140; commons-use, 121; community, 24, 59; corporate, 59; economic, 47, 103; effective, 106; efficient, 103; financial, 101; fiscal, 42; modern utility, 108; political, 47; private, 122;

professional, 39; service, 63; shifted, 136; sustainable, 108; sustainable environment, 107
management contracts, 92, 100, 126
management fee, 91, 97, 101; fixed, 101, 105, 125
management practices, 8, 140; bad, 103; best, 111; public–private, 106; unequal social water, 25
Manor, J., 23, 33, 41
Marcus, G.E., 7
market, 4, 22, 33, 40–1, 53, 68–9, 96, 121, 140; free, 5, 95, 103; internal, 97; open, 121; private, 119; self-regulating, 96
market access, 68
market competition, 53
market economies, 36
market forces, 4, 29, 32, 37, 97, 103
market mechanisms, 1, 55
market orientation, 6
market process, 34
market rules, 132
Mathur, K., 21, 23
McCully, Patrick, 25
McGrew, A., 2
MCM (million cubic meters), 76
MDGs (Millennium Development Goals), 61, 107
Medha Patkar, 138
Mehta Lyla, 25, 28
Mehdudia, S., 88, 116, 132
Menon, 66
meters, 77, 108, 116, 128; installing, 130; million cubic, 76; shared, 101, 110, 116
MGD, 77–8, 82–3, 85, 87, 96, 114
Millennium Development Goals *see* MDGs
million cubic meters *see* MCM
Mississippi River, 123
Missouri-Mississippi River System, 96
MNCs (Multinational Corporations), 7, 41, 67–9, 89–91, 93, 95, 103, 113, 120, 123–4, 126, 132–4, 137, 140, 143–4, 146
models, 4, 10, 14, 43, 102, 137, 146, 150, 152–3; applying borrowed, 135; bargaining, 150; centered, 149;

models (*continued*)
 classic, 11; classical, 2, 11, 26;
 horizontal, 149, 152; ideas and
 power, 30; mechanistic, 14; post-
 independence, 82
Mohan, Rakesh, 36–7, 51, 107, 109,
 124, 146
Mohanty, N., 64
Mudgal, Chitra 119
Mukherjee, Pranab 32
Mollinga, P.P., 6, 23–5, 153
Mooij, J., 9, 13, 20, 23, 25, 27, 149
movements, 97, 118, 120; effective
 entrepreneurial, 32; nationwide, 89;
 people's, 89, 94
Multinational Corporations *see* MNCs
multipurpose projects, 49
multiscalar, 6, 7, 141

Naqvi, 113
Narain, Sunita, 24, 59, 154
narratives, 20, 99, 114, 117–18, 137,
 140, 142–5, 153; competing, 152–3;
 mainstream, 142; shifting, 96
narratives of water, 27, 140
Narsalay, R., 43, 50
Narsimha Rao, 35, 38, 67
National Advisory Council (NAC), 119,
 123
National Advisory Council for Delhi,
 132
National Building Construction
 Corporation (NBCC), 85–6
National Capital Territory *see* NCT
National Federation of Indian Women,
 98
National Front Government, 38
National Urban Infrastructure Project,
 57
National Water Community, 60
National Water Policy (NWP), 43, 45,
 48–50, 52, 54, 58–60, 106–7, 142,
 146
National Water Resource Council
 (NWRC), 43, 49, 53, 59, 60
National Water Resources Policy, 53
Navdanya, 93
Nayar, B.R., 30, 33, 40
Nayyar, 22, 42

NBCC (National Building Construction
 Corporation), 85–6
NCT (National Capital Territory), 72–4,
 100, 134, 139
Nehru, Jawaharlal 27, 30
Nehru–Mahanalobian 29
Nehruvian, 42, 47
neoclassical, 4
neoimperialism, 127
networks, 6, 13, 15, 24, 33, 35, 137,
 143–6, 150; communicative,
 18; complex, 8, 152; core, 144;
 extended, 65; formal distribution,
 128; industrial, 112; international
 benchmark, 96; knowledge power,
 1; maintaining infrastructure, 103;
 public services, 54; road, 96
networks of power, 143–4
networks of resistance, 145
NGO Parivartan, 97, 125, 140
Nigam, A., 71
Nonrevenue, 81
nonrevenue water, 79, 96, 104, 123, 128;
 excessive, 111; reducing, 128
Nustad, K., 99

Old Tehri Town, 84
Ondeo, 108
Ondeo Degremont, 89, 94, 108, 113,
 119, 124–5, 145, 151
operating zone input, 125
operation and maintenance (O&M), 92
Operation Evaluation Study, 56
operational zones (OZ), 92, 100
operations, 52, 54, 88–9, 92, 100–1,
 108, 126; efficient, 109; fixed, 125;
 inefficient, 104; retail, 52; upgrade
 DJB's, 101
operators, 100–1, 105, 110; private, 52,
 91, 100, 129, 131; professional, 92;
 telecommunications, 42
opponents, 33, 93, 96, 110, 112, 122,
 128–9, 143
organizations, 11, 14, 19, 36, 85,
 95, 97; based, 63; consumer, 95,
 113; environmental, 93; farmer,
 93; independent civil society, 147;
 intergovernmental, 3; national, 95;
 neoliberal, 35; nongovernmental, 6,

21, 57, 71, 118, 139; religious, 95; voluntary, 63; women's, 93, 145

Paani Morcha, 95, 97
Painter, D., 69
Panchayats, 49
participation, 60, 105, 109, 118, 133, 142, 147; increased, 65; invited, 114, 140; private, 37, 62; stakeholder, 104–5
Pasteur, K., 26
Patnaik, P., 22, 39
Pederson, D.J., 39
penalties, 89, 91, 100–1, 105, 116, 125
people centered approaches, 142
per capita income, 30
performance: economic, 30; improved, 103; improved sector, 57
performance criteria, clear, 105
perspectives, 10, 13, 21, 26, 138; alternative development, 25; anti-state, 25; constructivist, 16; dissenting, 112; economic, 103; global, 35; historical, 29; incrementalist, 13; instrumentalist, 12; national, 50; reformist, 16
petitions, 97
planning: centralized, 30, 153; centralized development, 29, 43; centralized policy, 5; directed state, 5; water resource, 106
Planning Commission, 36–7, 61–3, 65, 104–5, 112, 119
Planning Department, 71, 74, 81, 83
players, 1, 3, 16, 72, 108, 139, 153; corporate, 114; external, 147; global, 135; international, 149
policy: coherent, 39; defined, 60; developing, 52; final, 60
policy actors, 13
policy advisors, 33
policy agenda, 14, 150; malleable, 14
policy analysis, 17, 20, 149–50
policy approach, 16; linear, 9
policy as prescription approach, 2
policy belief system, 15
policy change, 11, 14–16, 19, 22, 27, 72, 107, 115, 149
policy coalitions, 15, 143

policy communities, 12–15, 150
policy design, 11, 65, 75; local, 147
policy discourses, 1, 17, 22, 26, 51, 99, 142, 147, 150; competing, 71
policy documents, 24, 58, 63, 65–6, 99, 104, 122, 127; draft, 40; state water, 24
policy elites, 6, 146, 153
policy entrepreneurs, 13, 29, 33–5, 40, 50
policy formulation, 11, 23, 115, 118, 129, 149
policy instruments, 115, 143; advocacy of particular, 99, 143
policy language, 17
policy makers, 2, 12–13, 27, 49, 55, 57, 66, 98, 143, 150
policy networks, 12–13, 150, 154
policy process literature, 9, 12, 17, 23, 149
policy processes, 1, 6–7, 9, 13–18, 20–3, 25–9, 33, 60, 96, 99, 104, 117–18, 137, 141–2, 145–51, 153; agency frame, 27; deliberative, 20; domestic, 22; drive, 99, 150; dynamics of, 146, 150; dynamics of water, 6; emergent dynamics of, 21; environmental, 12; formal, 26, 144; influence, 14; political, 21; shaping, 117; subnational, 152; unexplored, 22; world, 20
policy production, 1–2, 5, 7, 66, 120, 151–2; national water, 60
policy reform, 57, 96, 115, 142; comprehensive water, 87; interests shaping, 33; situates contemporary water, 29
policy spaces, 12, 20, 33, 51, 68, 115, 117–18, 120, 147, 151–2
politics, 2, 6, 12, 21–6, 29, 66–7, 153–4; bureaucratic, 13, 149; corrupt patronage, 146; democratic, 144; domestic, 41; mass, 146; technology of, 27
pollution, 75, 97
population, 29, 41, 64, 71–2, 75, 77, 79–81, 121
poverty alleviation, 30, 54, 58, 66, 103, 107, 111, 114, 140, 142

Poverty Outreach Unit, 101, 111
poverty reduction, 69
power, 2–3, 5, 8–10, 13–20, 23, 25–7,
    29–30, 32–3, 39, 41–2, 49, 67, 69,
    139–40, 142–7, 151–3; centralize,
    132; class, 22; corridors of, 32, 114,
    144; discursive, 18; economic, 5;
    functional, 107; losing, 38; modern,
    27; social, 24; structural, 66
power differences, 7
power model, 34, 40
PPIAF (Public-Private Infrastructure
    Advisory Facility), 55, 57–8
PPPs (Public private partnerships), 51,
    55, 61–3, 65, 68–9, 89, 100, 103–6,
    111, 122–3, 128, 140, 143–4
practices: global, 127; independent, 18,
    104; local, 141, 144–5; micro, 19;
    political, 43; religious, 134
pressures, 4, 8, 40, 46, 77, 90–1, 95,
    127, 152–3; applied, 41; external, 4,
    58; global, 1, 3, 8, 152; growing, 49;
    internal, 4, 31; irregular, 79; low, 62;
    partisan, 38
Prime Minister, 30, 33, 36, 38, 62, 79,
    94, 123
Prime Minister Indira Gandhi, 30
Prime Minister Man Mohan Singh, 36
Prime Minister Narsimha Rao, 40
Prime Minister Rajiv Gandhi, 29, 49
Prime Minister Vajpayee, 59–60
principles, 5–6, 18, 67, 88, 107, 125,
    130, 138; basic, 132; commercial,
    103; economic, 37
private players, 51, 60, 64, 90, 92, 105,
    115, 121–2, 127–8, 140, 142; entry
    of, 51, 64; informal, 51; international,
    122
privatization, 5, 38, 41–2, 50–3, 61,
    65–6, 89, 94–5, 100, 108–9, 113–14,
    119–21, 123–4, 129, 132, 134–6;
    effects, 95; total, 92, 94
pro-poor measures, 110
process, 1–3, 5, 7–11, 13–22, 26–7,
    29–46, 50, 65–6, 69, 95–7, 113,
    119–20, 130, 138–9, 142–3,
    147–53
programs, structural adjustment, 4, 38,
    40, 43, 66, 68

protests, 33, 85, 90, 94–5, 97, 113, 120;
    mass, 141; public, 95, 113; staged, 85
public-private partnerships, 62, 100,
    115, 122, 135, 137, 140
Public private partnerships *see* PPPs
public-public partnership, 113, 132,
    135–6; demanded, 89
Pulsator Clarifiers and Aquazur, 88
PwC, 90, 92–4, 97, 99, 111–12, 119,
    127–8, 133, 139–40, 143–4, 147

quality of water, 54, 115, 123, 140; bad,
    115, 140; potable, 11

Raghu, 92
Rajamani, M., 61
Raju, 25
Ramachnandran, 154
Ramesh, M., 10, 13
Rao, 23, 37–8; Narasimha 41
Rashtriya Jal Biradari, 60
rationale, 57, 103; political, 66
Ratnagar, Shareen 70
Ravindran, P., 64
raw sewage, 122; dumped, 123
reality, 14, 16–17, 19, 26, 32, 124,
    128–9; frames, 19; ideological,
    50; political, 50; social, 17, 122;
    sociocultural, 152
reform process, 22, 35–7, 42–3, 51–2,
    61, 82, 96, 104, 132
reforms, 7, 20, 22, 27, 37–8, 40–3, 50,
    54–5, 57, 66–7, 69, 74–5, 95, 100–1,
    104, 110; basic, 97; complete, 33;
    decisive, 33; first generation of,
    35, 50–1, 66; fiscal, 67; generation
    of, 35–7, 42, 58; hard, 41; internal
    administrative, 42; legal, 51;
    motivated, 4; pass, 40; public sector,
    55; public sector expenditure, 68;
    push policy, 34; soft, 41; soft sector,
    43; suggested, 143; undertaking, 105;
    water utility, 63
relations, 1, 8, 18, 20, 24, 27, 42, 152;
    center–state, 22, 70; centre–state, 22;
    international, 3
religion, 133, 135, 141
religious groups, 93, 137, 145
reservoirs, 45; natural, 113

resident water associations, 93, 145
Resident Welfare Associations (RWAs), 93, 97, 104, 113, 119, 135–6
resources, 11, 13–15, 30, 36, 49, 59, 61, 65, 76, 121, 123, 125, 129, 132, 146; common property, 25; managing, 45; national, 50; natural, 25, 43, 46, 49, 51–3, 127
revenues, 46–7, 67–8, 70, 81, 92, 96, 114, 135; generated, 123; increased, 92; low, 123; maximize, 46–7; produced generated cash, 123; reduced, 67; reducing, 67; tax, 67
rights, 18, 25, 43, 51, 65, 68, 97–8, 118, 129, 133, 142; constitutional, 137; definite property, 46; democratic, 131, 137; exact, 59; land tenure, 128; tenured land, 116; transient property, 121
river Yamuna, 114
rivers, sacred, 89, 134
Roe, 12, 146
Roy Aruna, 119, 131
Roy Arundhati, 123
Rogers, 14
Rosenau, J.N., 3
Rudolph, L., 23
Ruet, J., 79

Sabatier, P., 11, 15–16, 19, 27, 148
Samans, Richard, 63
SANDRP, 59
Sanjay Sharma, 113
Sassen, S., 3
Satya, Laxman, 46
Saur, 142
scarcity, 28, 58, 100, 114, 140; forced, 120
Schaffer, 12, 14
schemes, 68, 122; cooperative, 136; linear, 12; multivalent, 2; sponsored, 67; water resource, 59
Schon, D., 13, 17
Schonwalder, G., 118
Schram, S., 17
science, 19, 24, 59, 94, 97, 112, 115, 130, 140, 142, 144, 153–4; neutral language of, 18, 27; political, 3, 21
Scoones, I., 10, 12, 14, 20–1, 33

sectors, 4, 8, 25, 36, 55, 57, 60, 62, 68, 104, 112, 152; corporate, 4, 63; industrial, 30; insurance, 43; marginalized, 69; public, 32, 46–8, 51–2, 67, 122, 142; sewerage, 111, 126; strategic, 66
Sehgal, R., 88–9, 95, 119
service delivery, 4, 53–4, 56, 73
service providers, 57, 69, 103; accountable, 7, 69, 111
Seventh Schedule, 48
sewage disposal, 49
Shackley, S., 12
Sharma, S., 25, 79, 113–14, 123
Shastri, Vandana, 22, 33–5, 67
Shiva, Vandana, 46, 68, 82, 88–9, 120–2, 124, 135–6, 140
Simon, H.A., 11
Singh, A.K., 38–9, 55, 60, 79
slums, 81, 110, 131
social groups, 41; marginalized, 118
social justice, 30, 50, 60
social movements, 3, 25, 115, 118–19
spaces, 6, 13, 20, 26–7, 29, 31, 40, 43, 59, 99, 115, 117–19, 137, 147–8; circumscribing, 150; invited, 117–18, 142; political, 141, 150
Sridhar, G., 25
stakeholders, 27, 55, 57, 59, 104–5, 117–18, 147; invited, 115; uninvited, 115
state electricity boards (SEBs), 43
state Governments, 5, 22, 29, 43, 49, 67, 69, 114, 140; compelled, 67
states, 1–8, 14–15, 22–4, 27, 29, 31, 39–43, 47–9, 64–70, 96, 117–18, 120, 135–6, 142–3, 145–9, 151; centralized, 47; colonial, 47, 135; compelled, 64; corporate, 132; independent, 43; independent Indian, 5; interventionist, 32; like-minded, 58; nation, 3; neoliberalizing, 121; oil-producing, 30; post-colonial, 135; post-colonial Indian, 47; pressure, 67; report, 104; set India's, 42; strong, 5; undemocratic, 132; weak, 3; world class cyber, 72
street level bureaucrat, 11, 14
subnational, 1, 7–8, 146, 152

subnational dynamics, 66, 147
subnational levels, 58, 68, 114
subsidiary, 88
subsidies, 32, 40–1, 55–7, 60, 68–9, 96, 103, 116, 124, 128, 130; direct, 129; operating, 92; transparent, 106
subsidized electricity, 88
subsistence, 133
Suez, 83, 122, 124–6, 133–4, 137, 142
suppliers, 81; effective, 52; formal water, 25
supply, 36, 51–2, 54, 76–80, 87, 89, 96–7, 112, 122, 129, 131, 136; bulk, 104, 107; continuous, 108, 112; enhanced, 146; intermittent, 106; irregular, 112; regular, 100; regularize, 108; regularized, 107; round-the-clock, 119
sustainability, 60, 65, 92, 100, 103, 121; ecological, 66; environmental, 64, 111–12, 115, 141; physical, 59
sustainable development, 61, 65, 103, 107, 142
Swadeshi, 36, 42
Swadeshi Liberalism, 42
system, 3, 6, 15, 22, 32, 45, 51, 81, 101–3, 108–10, 114, 142; administrative, 4; belief, 134; complex industrial licensing, 43; drainage, 74; exploitative land revenue, 46; federal, 5; political, 9, 49; public, 136; quasi-federal, 22; rapid transit, 72; solid waste management, 61; traditional knowledge, 24

tanks, 24, 35, 45, 57
targets: annual, 91; loss reduction, 129
tariff hikes, 93, 95, 113, 123, 134; claimed, 141; tenfold, 94, 113
tariff structure, 92; raised, 113, 136
tariffs, 42, 56–7, 93, 100, 102, 110, 123–4, 126, 130, 135–7; depoliticization of, 127; low, 109; official, 102; setting, 101; standard, 51
Tarun Bharat Sangh, 95, 97

technical approaches, 96, 141, 143
technology, 2, 6, 39, 63–5, 74, 88–9, 94, 107–9, 112, 115, 122, 130, 140–4, 153; modern, 39, 105, 108; plant uses, 88; political, 18–19, 109; state of the art, 88
Tehri, 83
Tehri Dam, 48, 83–5, 88, 97, 124, 145
Thakkar, Himanshu, 60
Thapar Romila, 70, 119
Thomas, 12, 20
Thompson, A., 11
trade, 53, 70; free, 70; industrial, 36; international, 6, 72; liberalize, 53
transmission losses, 88
treated water, 52, 79, 87–9, 107, 122
treatment capacity, 111; combined, 77
treatment plants, 110–11, 114, 116, 119; subsidized water, 131

ULBs (Urban Local Bodies), 49, 52, 61–2, 107
UN Covenant on Human Rights, 138
Union List, 42, 48
Union of India, 67
United Front Coalition, 42
United Provinces, 46–7
Universal Declaration of Human Rights, 138
Upper Ganga Canal, 77, 85
Upreti, B.C., 24
Urban Local Bodies *see* ULBs
urban water policy reform, 134
urban water reform project, 117, 122
urban water supplies, 49, 61, 65, 96, 114, 119, 127–8
user charges, 92
users, commercial, 136
utilities, 58, 96, 113, 127, 136

Vaidyanathan, A., 24
Venkataramanan, 35
Vihar, Sonia, 7, 69, 82–3, 87, 89, 94, 108, 114, 119, 124, 131, 137, 146–7, 152
vision, 24, 32, 35, 42, 104, 111–12, 114, 116, 119, 140–1, 146, 148; developmental, 112; developmentalist, 146; international,

107; neoliberal, 66; nuanced, 104; radical, 66; technocratic, 19, 47
Viswanathan, S., 121
voices, 7, 9, 20, 26, 33, 104, 118, 129, 137, 144–5, 148, 152; dissenting, 144, 147
Vombhatkere, S., 60

Wadsworth, 154
Walton, 154
wastewater, recycling, 113; discharge, 73
water, 4–9, 23–4, 26–9, 43, 45–70, 75–9, 81–2, 84–91, 93–7, 99–100, 102–6, 108–16, 118–23, 125–38, 140–4, 152–4; accessing, 128; brown-colored, 122; clean, 102; constructions of, 99, 104; cost of, 49, 65, 124; defined, 53; distribute, 125; diversion of, 85; divert, 129, 131; drink, 135; fetch, 110; frame, 142; free, 116, 128, 135; ground, 89; illegal, 116, 128; improving, 111; managing, 49; piped, 110, 129; post-reform era, 142; potable, 73, 77, 103, 109–10; privatize, 112, 128; reallocate, 69; reclaim, 133; represented, 115; resident steals, 136; respective municipal, 69; revealed, 139; rightful custodian of, 59; sacred, 133–4; safe, 7, 107, 121; safer, 102, 111; the scarcity value of, 54; share, 120; storm, 74; supplied, 142; supplying, 81, 116; surface, 76–7; surrounding, 123; tap, 131; transferred, 145; turbulent, 64; untreated, 81
water access, 63
water activists, 133
water apartheid policy, 133
water availability, 76, 79
water boards, 49
Water Campaign, 95
water charges, 54, 60, 109
water conservation, 51
water crisis, 64–5, 121; growing, 97
water culture, 132
water demand, 73, 77, 81; managing, 110

water democracy, 93–4, 113, 119, 132–3, 137; country's, 132
water distribution, 47, 61, 91, 93, 100, 103, 109, 121–2; modernize Delhi's, 112; privatize, 119
water governance, 7, 151
water harvesting, 24–5, 154
water infrastructures, 64, 142
water knowledge, 26, 99, 151
water leakages, 135
water legislation, 94
Water Liberation Campaign *see* WLC
Water Man, 95, 135
water management, 5, 28, 45, 47–9, 59, 72, 96, 100, 103; efficient, 54; improved, 54
water markets, 50, 135, 137
water per capita, 79
water policy, 1–154; desirable, 6, 23–4; draft National, 53, 58; pressures shaping, 139; producing, 7; prudent, 121; shaping, 117; state, 60, 106; urban, 7, 27, 45
water policy processes, 2, 6, 9, 25–6, 139–53
water policy reform, 43, 45, 50, 58, 66, 68, 109, 115, 130, 143, 146
Water Policy Reform Initiative, 55, 57
water policy reform knowledge, 150
Water Politics, 23
water pollution, 49
water power, 48
water prices, 93, 109
water privatization, 7, 89, 94, 110, 119–20, 123, 132, 137; fight, 97–8; prioritize, 136; stop Delhi's, 94
water reform policies, 66, 99, 115, 118, 143, 147, 152; urban, 26, 153
water reform project, 71–9, 81–2, 85, 87–93, 95–8, 117, 119, 122, 137
water regulator, 130
water requirements, 88
water resource development, 49, 59, 119
water resources, 24–5, 43, 47–9, 55, 59, 63, 65, 71, 77, 97, 103, 106, 112, 127, 134, 151; city's, 71; natural, 73; precious, 136; scarce, 128; sustainability of, 66, 132

water resources management, 4, 6, 12, 23–5, 49, 54–56, 59, 103, 127; sustainable, 54
water rights, 43, 89, 110, 120, 132; transferable, 69
water scarcity, 25, 48, 65, 69
water sector, 6, 37, 50, 52–3, 55, 57, 63–4, 66, 69, 75, 89, 96, 100, 108, 114, 140; controversial, 41
water security, 47
water stakeholders, 152
water supplies, 36, 41, 48–9, 51–3, 61–2, 64–5, 67–8, 77–8, 90, 100–2, 105–6, 108–9, 111, 115, 125–6, 136; allocated urban, 48; augmentation of, 83; augmenting, 114; bulk, 116; city's, 122; continuous, 101, 106–7, 111; distributed, 100; improving, 82; inadequate, 79–80, 104, 106; insufficient, 115, 140; insufficient drinking, 48; intermittent, 78, 111; non potable, 140; piped, 79; privatize drinking, 105; privatizing, 96; public, 138; restructuring, 91; rural, 49, 77; safe, 102, 111; total, 76; treated, 148; uninterrupted, 91
water supply boards, 52
water supply problems, 142–3
water tankers, 81, 102, 116
water tariffs, 7, 93–4, 109, 124, 133; increasing, 52; low, 51; lowest, 109; pre-revised, 109
water treatment plants, 7, 68, 76–7, 79, 82–3, 113
Water Workers Alliance *see* WWA

Waterbury, J., 33
waterlogging, 49
watershed management, 27, 63
Watkins, 96
Weber, 11
WEF (World Economic Forum), 63
WEF Water Initiative, 63
Weiner, M., 22, 70
Weiss, Linda, 3, 12
welfare, 37, 51, 67, 97–8
Western Yamuna Canal, 77
Williamson, John, 37
Wilson, D.11
WLC (Water Liberation Campaign), 89, 94–5, 97, 113, 118, 120–2, 124, 126–30, 132–5, 139, 143–5, 147, 153
World Bank, 32–3, 35–41, 53–6, 58, 63–6, 68–9, 90, 94–5, 97, 99–105, 112–15, 127–9, 132–3, 139–40, 143–4, 146–8
WWA (Water Workers Alliance), 93–5, 104, 112–13, 130, 136, 147

Yamuna, 74–5
Yamuna Basin, 113
Yamuna River, 73–4, 76–7, 136
Yamuna River Chandrawal, 78
Yamuna River Haiderpur, 78
Yamuna River Wazirabad, 78

Zerah, M.H., 79, 81
zones, 90–3, 97, 101, 103, 110–11, 126; input, 87, 116; operating, 90–1, 125; operational, 92, 100

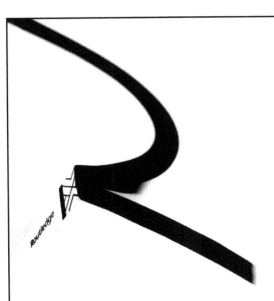

Milton Keynes UK
Ingram Content Group UK Ltd.
UKHW040057071024
449327UK00019B/622